T0259710

Environmental Footprints and Eco-design of Products and Processes

Series editor

Subramanian Senthilkannan Muthu, SgT Group and API,
Hong Kong, Hong Kong

This series aims to broadly cover all the aspects related to environmental assessment of products, development of environmental and ecological indicators and eco-design of various products and processes. Below are the areas fall under the aims and scope of this series, but not limited to: Environmental Life Cycle Assessment; Social Life Cycle Assessment; Organizational and Product Carbon Footprints; Ecological, Energy and Water Footprints; Life cycle costing; Environmental and sustainable indicators; Environmental impact assessment methods and tools; Eco-design (sustainable design) aspects and tools; Biodegradation studies; Recycling; Solid waste management; Environmental and social audits; Green Purchasing and tools; Product environmental footprints; Environmental management standards and regulations; Eco-labels; Green Claims and green washing; Assessment of sustainability aspects.

More information about this series at http://www.springer.com/series/13340

Subramanian Senthilkannan Muthu
Editor

Environmental Water Footprints

Concepts and Case Studies from the Food
Sector

 Springer

Editor
Subramanian Senthilkannan Muthu
SgT Group and API
Hong Kong, Hong Kong

ISSN 2345-7651   ISSN 2345-766X (electronic)
Environmental Footprints and Eco-design of Products and Processes
ISBN 978-981-13-4777-1   ISBN 978-981-13-2454-3 (eBook)
https://doi.org/10.1007/978-981-13-2454-3

© Springer Nature Singapore Pte Ltd. 2019
Softcover re-print of the Hardcover 1st edition 2019
This work is subject to copyright. All rights are reserved by the Publisher, whether the whole or part
of the material is concerned, specifically the rights of translation, reprinting, reuse of illustrations,
recitation, broadcasting, reproduction on microfilms or in any other physical way, and transmission
or information storage and retrieval, electronic adaptation, computer software, or by similar or dissimilar
methodology now known or hereafter developed.
The use of general descriptive names, registered names, trademarks, service marks, etc. in this
publication does not imply, even in the absence of a specific statement, that such names are exempt from
the relevant protective laws and regulations and therefore free for general use.
The publisher, the authors and the editors are safe to assume that the advice and information in this
book are believed to be true and accurate at the date of publication. Neither the publisher nor the
authors or the editors give a warranty, express or implied, with respect to the material contained herein or
for any errors or omissions that may have been made. The publisher remains neutral with regard to
jurisdictional claims in published maps and institutional affiliations.

This Springer imprint is published by the registered company Springer Nature Singapore Pte Ltd.
The registered company address is: 152 Beach Road, #21-01/04 Gateway East, Singapore 189721,
Singapore

This book is dedicated to:
The lotus feet of my beloved Lord
Pazhaniandavar
My beloved late Father
My beloved Mother
My beloved Wife Karpagam and
Daughters—Anu and Karthika
My beloved Brother
Everyone working in the food sector to make
it ENVIRONMENTALLY SUSTAINABLE

Contents

Environmental Footprints of Water—Concepts, Tools, Importance and Challenges

P. Senthil Kumar and K. Grace Pavithra

Abstract The worldwide demand for clean water makes water a vital importance in supply and efficiency in usage for the sustainable future. Rapid industrialization and economy, increases water demand mainly in the field of agriculture and industrial sector. There is vulnerability for the available quality of water due to the climate variability and raising demand. In order to predict the demand of water, footprint assessment techniques and tools are introduced in monitoring greenhouse gases and water flow across the world in last decades. This chapter provides a detail sketch of green, blue, grey water, virtual water and its global trends. The detailed review of water management in energy sectors such as, integration of waste water with water management planning, improvement in cooling systems, development and integration of decision-support tool with weather models and climate, their importance as well as future challenges are explained in detail.

Keywords Demand · Industrialization · Vulnerability · Water Wastewater integration · Assessment tool · Footprint assessment techniques

1 Introduction

The overall fresh water availability is 2.5% and among 2.5%, 68.1% are in the form of ice, 30.1% are in the form of 30.1 and 1.2% of surface water. The fresh water availability in domestic sector is 11%, 19% in industry and finally in agriculture 70% (Bhat 2014). With reference to time and space precipitation is found to be renewable and the pathways are considered as green and blue water flows. The distribution of freshwater across the world is uneven and it is found to be essential element for humans and ecosystems. The fresh water availability for human consumption is under vulnerability due to climate, water supply and water demand and competition of freshwater resources are seen in past decades due to increase in

P. Senthil Kumar (✉) · K. Grace Pavithra
Department of Chemical Engineering, SSN College of Engineering, Chennai 603110, India
e-mail: senthilchem8582@gmail.com

© Springer Nature Singapore Pte Ltd. 2019
S. S. Muthu (ed.), *Environmental Water Footprints*,
Environmental Footprints and Eco-design of Products and Processes,
https://doi.org/10.1007/978-981-13-2454-3_1

population, economic growth, demand for agricultural products, industrialization and energy production. The surface water supply is associated with uncertainties with droughts and distribution of rainfall. In 2017, UN World Water Development Reported on wastewater that, the demand of water as well as the volume of wastewater produced increasing continuously. Around 80% of world's wastewater 95% are found in some developed countries which are released without treatment. The water once produced either diluted into river, lake, streams or transported to downstream. With referred to time and space precipitation is found to be renewable and the pathways are considered as green and blue water flows (Schneider 2013). Green water is considered as purest form of water and seen in the form of soil moisture, used by plants via transpiration. Surface and ground water which are stored in lakes, stream groundwater, in the form of glaciers and snow are considered as blue water (Rodriques et al. 2014). Grey water (a product water of domestic activities and not in contact with fecal matter) and black water (sewage water flushed in the toilets) are the transformation of blue water which is in polluted form. The quantity of water utilized in food and in other products are referred as virtual water. Due to change in climate, limited water supply and demand, fresh water availability for human consumption is under threat. The uncertainty followed in spatio-temporal distribution of rainfall as well as multi-year droughts makes complications in the surface water supply. In future, due to impact of socioeconomic and drought there will be complexity in freshwater availability. There is continuation in the demand for water due to population growth, industrialization, agriculture, domestic use etc. (Vorosmarty et al. 2010; Srinivasan et al. 2013). It is expected that by 2025 around 1.8 billion people will witness water scarcity (WWAP 2012; WWDR 2015) and the percentage of water consumption for energy and agriculture production will increase by 2035 (IEA 2011). In order to attain sustainable management, water availability and its vulnerability in a changing environment are to be quantified.

This chapter provides detailed discussion about the water footprint and its components with its usage in primary sectors. Water footprint in terms of environmental sustainability were discussed. The water footprint general process steps in various primary sectors like, agriculture and forestry, wastewater treatment plants, some of the manufacturing industries like, textile, paper, food and beverages were included and finally the chapter has been concluded with establishment of WF benchmarks and challenges faced with the incorporation of water footprint in primary sectors.

Water Footprint-Introduction

The water footprint is a measure of humanity's abduction of fresh water in volumes either consumed or polluted. In other terms it can be represented as the amount of water utilized for each services and goods we use. From single process like growing of rice, multi-national company to a particular country from an aquifer or river basin. Water footprints answers the questions for companies, governments and individuals regarding water dependence in company's operation, water resources regular protection, security of food or energy supplies etc. (McKinsey 2009).

The Water Footprint Concept

Freshwater is one of the most valuable assets and it is becoming progressively rare. There is a need to evaluate how much of freshwater is accessible what's more, its human appointment over a specific period. The water footprint communicates the human assignment of freshwater in volume. The comparison of human's water footprint with freshwater availability is considered to be part of water footprint sustainability assessment. Today, one out of 10 individuals on the earth not have clean water for accessing and one in each three individuals don't have access to water for sanitary purposes. Water table level is diminishing at a speedier rate than it can be renewed (Strauss 2016). The water resources are over exploited due to human activities. In environmental agendas of nations, companies, decision makers and the public as well as across the media, the water scarcity has become an important issue. There is a need to conserve our water resources and protect our ecosystem by reducing water footprints (Ercin et al. 2012). The concept of water footprint was introduced in 2002 by Arjen Y. Hoekstra for measuring the used water. A water footprint refers to the amount of freshwater used directly as well as indirectly by an individual, community or country for a period of time can be represented in scientific term as water footprint. It is a product of fresh water volume used to produce a product, which is measured throughout the supply chain. It is referred as multidimensional indicator, showing water consumption volume and polluted water volume by source and type of pollution. (WATER FOOT PRINT 2). In 2011 Hoekstra stated four-step approach for fresh water appropriation, which is shown in Fig. 1.1.

Water Footprint Components (Water 2017)

Three components of water footprint namely, green water footprint, blue water footprint and grey water footprint. The first two account for total consumption and the last one determines the amount of fresh water polluted. The quantity of water used is calculated in terms of water footprint. It indicates amount of water consumed and contaminated during industrial processes. Figure 1.2 shows the components of water footprint.

Green water footprint—Rainwater is referred as green water footprint which does not run off or recharge the groundwater but stored in soil or stays on stop of soil. It refers to the rain water volume consumed for the production of various agricultural and forest products. It is a summation of water lost in evapotranspiration and precipitation and the amount of water locked in harvest.

Blue water footprint—It refers to the amount of groundwater or surface water consumed along the supply chain of a product or service. Domestic water use, industry and irrigated agriculture water usage comes under blue water footprint.

Grey water footprint—Grey water footprint is an indicator of amount of pollution in freshwater with the production of a product over its full supply chain. It is calculated based on the volume of water requires to dilute pollutants to an extent to meet water quality standards.

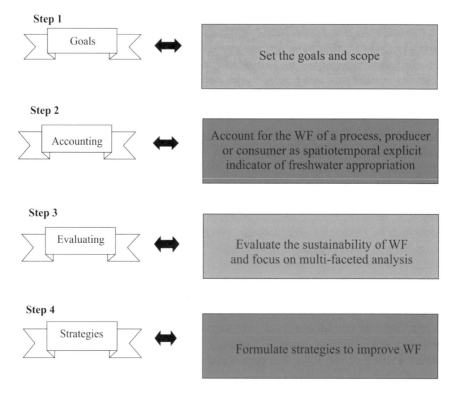

Fig. 1.1 Key components of water footprint

Water Footprint Standards

It is an indicator or a set of indicators which reflects the impact of human activity over to our environment. ISO developed water footprint standard and there is no universal followed internationally. ISO14046: 2014 specifies principles, requirements and guidelines, which are related to water footprint assessment of products, processes and organizations based on life cycle assessment (LCA). LCA assess the stages of the product from cradle to grave. It provides guidance and requirements for calculating and reporting water footprint assessment (ISO 2014).

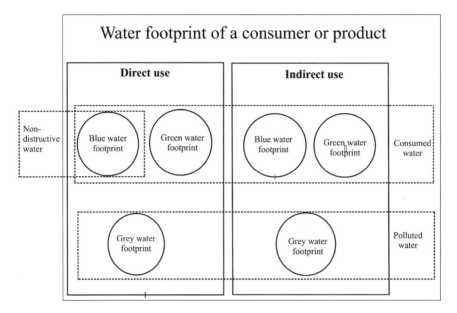

Fig. 1.2 Water footprint of a consumer or product

ISO suggest that the organizations to use water footprint to:

- Assessment of magnitude of environmental impacts which are related to water.
- Development of risk management related to water.
- Optimization of water management at product, process and organizational levels.
- Provision of scientific results as evidence in reporting water footprint results.
- The global water footprint assessment standard has been accepted worldwide for conducting water footprint assessment.

  The instructions and guidance are as follows,

- Calculation of green, blue and grey water footprint by understanding the geographic and temporal allocation of water resources for agriculture, industry and domestic water supply.
- Conducting water footprint sustainability assessment by understanding the environmental sustainability, resource efficiency and social equity.
- Usage of the results in water footprint accounting and sustainability assessment to identify and the actions to be taken in local, regional, national and global scales.

On the whole, the global water footprint assessment standard provides us the way to achieve smart usage of world's fresh water. As the global water footprint assessment standard provide quantification and robust analytics of water which support companies on corporate water sustainability journey. It can be used for different geographic scales irrespective of the quantity of water availability. The standards can be used by the researchers to develop water footprint statistics.

2 Water Footprint as Environmental Sustainability

For the development of human activity, freshwater is found to be natural resource key which is essential for human surviving and wellbeing. In most economic sectors as well as environmental asset, water is found to be one of the factor for production and it is one of our right to access clean water and sanitation as per sustainable developmental goals (United Nations 2010; Martinez-Paz et al. 2014). Comparing to water extractions and the pollution, the renewal of natural water cycle was found to be minimal in some places (Pophare et al. 2014). It is necessary to study and evaluate the current usage of water resources in order to predict the availability of quality as well as the quantity of water resources. Many index (such as Water Strex Index, Water Exploitation Index, Water Allocation Index, Exploitable Water Resources etc.) which explicit or implicitly tracks the sustainable use of water resources in a province and this assessment find its way in drought or water scarcity situations (Sandoval-Solis et al. 2011). The water footprint (WF) is found to be new indicator which quantifies the usage of fresh water in terms of production factor. Unlike other indicators it does not measure only the amount of water resources, the pollution of green water is also included (Lovarelli et al. 2016). From the adaptation of ISO 14046 norm, the WF has become the international reference for the impact assessment of the product, services, processes and organizations over water resources (Hoekstra 2016) and WF can also be used as an indicator for the geographical area (Dong et al. 2013). As WF gives total amount of water utilized in geographical area, it can be used as an indicator for water resources management. As river basin is considered as common spatial unit in water planning process, sustainability analysis should be done for river basin. Each component (WF_{Green}, WF_{Blue}, WF_{Grey}) is compared to its maximum values to maintain sustainability of river basin. In dealing with geographical area, sustainability of water footprint is aggregated in certain area, catchment area or river basin, as it is the natural unit in which WF is easily compared with water availability, water allocation and with conflicts. The contribution of WF for individual process, product, consumer or producer are taken into account. The contribution of water resources lies with two elements: (i) The contribution of water for specific process, product, consumer or producer, and (ii) The contribution of aggregated water footprints in specific geographical areas. The scope of environmental sustainability of WF depends upon the following checklist and the checklist are as follows,

- The sustainability of the green, blue and grey water footprint.
- Sustainability of environmental, social and economic dimensions.
- Identification of hotspots in detail with including primary or secondary impacts.

The betterment of hotspots lies with the spatial and temporal resolution level. For better understanding of WF in a geographical area, one needs to describe the water footprint of a catchment area affecting water flows and quality of water resources and its impact to the indicators (such as welfare, social equity, human health, biodiversity etc.). In a process, WF is found to be sustainable when it matches with following two criteria (Hoekstra et al. 2011a, b):

- Using current technologies, it is impossible to carry out a process at lower WF.
- The internal WF of a river basin is sustainable and there is no raise in environmental, social or economic issues.

A methodology was applied by Pellicer-Martinez (2016) to Segura River Basin (SRB), South-eastern Spain, which is considered as most complex river basins in Europe. The methodology starts with simulation of anthropized water cycle, by combining hydrological model and decision support system under several conditions. The simulations were analyzed in several temporal circumstances.

3 Water Footprint in a Process Step

3.1 Blue Water Footprint

The consumptive use of blue water is stated as blue water footprint as previously said. The 'consumptive water use' refers to the following scenarios:

- Evaporation of water
- Incorporation of water into a product
- The water which does not return to the same geographical area
- Water does not return in the same time when it is taken.

The blue water footprint in a process step can be calculated as

$$WF_{blue} = Blue\ water\ evaporation + Blue\ water\ incorporation \\ + Lost\ return\ flow[volume/time] \tag{1}$$

The return refers to the non-availability of water for reuse within the same catchment within the same period.

3.2 Green Water Footprint

Precipitation on land which does not run off or recharge the ground water but is stored in the soil or temporarily stays on the soil. The agricultural and forestry products which consumes rainwater are considered as green water footprint and the green water footprint for process are as follows,

$$WF_{proc,green} = Green\ water\ evaporation \\ + Green\ water\ incorporation[volume/time] \tag{2}$$

Using empirical formulas, green water consumption in agriculture can be estimated by estimating the evapotranspiration based upon the input data on climate, soil and crop characteristics.

3.3 Grey Water Footprint

It is the degree of freshwater pollution which is associated with process step. Calculation of grey water footprint can be done by dividing the pollutants load (L, in mass/time) by the difference between the ambient water quality standard for that pollutant (maximum concentration C_{max}, mass/volume) and its natural concentration (C_{nat}, mass/volume).

$$WF_{proc,grey} = \frac{L}{C_{max} - C_{nat}} [volume/time] \tag{3}$$

Natural concentration refers to the water body were no human disturbances in the catchment. In grey water footprint when C_{nat} is not equal to zero. The ambient water quality standards are to be followed. For particular substance, the water quality standard varies from one to another water bodies.

4 Water Footprint Tools in Various Primary Sectors

4.1 Water Footprint in Agricultural Sector and in Forestry

Products such as food, fibre, fuel, oils, soap, cosmetics etc., contains ingredients from agriculture or forestry. Agriculture and forestry sectors are major water consuming sectors has significant water footprint. The water footprint of the process of crops or trees (WF_{proc}) is the sum of green, blue and grey components:

$$WF_{proc} = WF_{proc,green} + WF_{proc,bue} + WF_{proc,grey} [volume/mass] \tag{4}$$

Usually, water footprints in agriculture or forestry is expressed as m^3/ton equivalent to litre/kg.

The water footprint blue component for a growing crop or tree can be calculated as the blue component in crop water use divided by the crop yield. Similarly, the green component can also be calculated in same pattern.

$$WF_{proc,green} = \frac{CWU_{green} \left(\frac{m^3}{ton} \right)}{Y \left(\frac{ton}{ha} \right)} \tag{5}$$

$$WF_{proc,blue} = \frac{CWU_{blue}\left(\frac{m^3}{ton}\right)}{Y\left(\frac{ton}{ha}\right)} \tag{6}$$

This method is applicable for both annual and perennial crops. For perennial crops, annual yield over its life span should be considered. By accounting the initial yield of planting equal to zero and yield attainment to the highest level after some times and the yield goes down at the life span of perennial crop. For estimation of crop water use, average annual crop water use over the crop's life span are to be accounted. The grey component water footprint of a growing crop or tree is calculated based upon the chemical rate involved per hectare (AR, kg/ha) times to the leaching run-off fraction (α) divided by the difference between the maximum acceptable concentration (C_{max}) and the natural concentration for the pollutant (C_{nat}) and then divided by the crop yield (Y).

$$WF_{proc,grey} = \frac{\frac{(\alpha X AR)}{(C_{max}-C_{nat})}}{Y} \; [volume/mass] \tag{7}$$

The pollutants are generally fertilizers (such as nitrogen. Phosphorus etc.), pesticides and insecticides. Waste flow to freshwater bodies has to be considered as pollutants and a fraction of total application.

The global estimate of the water consumption for several crops ins different countries are estimated by Hoekstra and Hung (2002). But the research does not split the composition into green, blue and grey water and numerous studies were carried out by many researchers and according to Chapagain and Hoekstra (2003) the crop production for the period 1997–2001 was 6390 Gm^3/year and the WF quantified for maize, wheat and rice are 900, 1300 and 3000 m^3/t respectively. Mekonnen and Hoekstra (2010) and Hoekstra et al. (2011a, b) studied WF for various products and processes assessing agriculture where sugar crops and vegetables showed low WF (200 and 300 m^3/t). Fruits and oil crops showed 1000 and 2400 m^3/t. Pulses, spices and nuts required higher quantities of WF, which varied between 4000 and 9000 m^3/t, respectively. Vanham and Bidoglio (2013) reviewed WF of EU28, results showed that Italy and Spain consumed two thirds of European irrigation (WF$_{blue}$). Italy, Spain, Portugal, Greece and France accounted for 965 of EU WF$_{blue}$ and France, Spain, Italy, Poland, Germany and Romany were consumers of WF$_{green}$. In many studies, WF assessment were done and three components green, blue and grey were highlighted. The features such as soil texture, pH, nutrient and mineral soil content as well as the market policies and the economic return are taken into account.

4.2   Water Footprint in Wastewater Treatment Plants

In WWTPs, the blue water footprint is accounted in each unit operations and in the evaporation of water during the treatment of wastewater. The green water footprint is not considered here as WWTPs does not utilize the water which is incorporated within the soil and the grey water footprint is calculated using WFA manual (Hoekstra et al. 2011a, b). For WWTPs, specific calculations are adopted and the calculations are based on mass balance which is absorbed at WWTP discharge point.

Mass balance at discharge point of WWTP of the pollutants

$$Q_e C_{e(p)} + WF_{grey} \cdot C_{nat(p)} = \left(Q_e + WF_{grey(p)} \right) \cdot C_{\max(P)} \tag{8}$$

Mass balance of pollutants using grey WF equation

$$WF_{grey} = max \left[WF_{grey(p)} \frac{\left(Q_e \left(C_{e(p)} - C_{\max(p)} \right) \right)}{\left(C_{\max(p)} - C_{nat(p)} \right)} \, volume/time \right] (for\, p = 1\, to\, p) \tag{9}$$

Q_e effluent flow rate (volume/time)
$C_{e(p)}$ concentration of a pollutant p in the WWTP effluent
$C_{\max(p)}$ concentration of pollutant p allowed in receiving water body
$C_{nat(p)}$ natural concentration of pollutant in receiving water body.

Grey WF are calculated for each compound because of the existence of many number of pollutants in WWTP discharge. A case study of La Garriga WWTP and the Congest river in Spain was taken. The La Garriga WWTP treats 4,000 m^3 d^{-1} and discharges it to Congost river. It was designed for 29,000 population which treats organic matter and nitrogen. The pollutants concentration such as total nitrogen, total phosphoris and total organic carbon are utilized for the calculation of $EF_{grey(p)}$ and WF_{blue}were applied to the water consumption sectors which were obtained from Ecoinvent 3.0 database. The C_{max} concentration information's of Besos river Basin was obtained from River Basin Management Plans from Catalonia and the C_{nat} were taken from the upstream of WWTP. A sensitivity analysis was done over the important factors which are contributed on the results by increasing and decreasing the factors such as concentration of phosphorus, maximum concentration of phosphorus permitted into the river and the electricity consumption of the plant. Since, they are found to be major contributors of water footprint. The water treatment which was under study reduced the water footprint by 51.55 in secondary treatment and 72.45 in chemical phosphorous removal. Since phosphorous removal consumes higher amount of water footprint, it should be considered among along the pollutants (Morera et al. 2016).

4.3   Water Footprint in Mining

Water is used in various steps in recovering valuable metals from its ore. For example, for obtaining 19 kg of copper which is found in medium sized family car, around 1600 L of water are needed. The movement of material in large scale and voids development due to mining leads to alteration in the flow paths of the water. Water which infiltrates into underground mines are elevated to the surface in order to avoid flooding of voids and mine shafts. Little amount of water is required in mining and only less amount of water is used in fire control, equipment cooling and in dust suppression. Two types of approaches are seen in mining, they are categorized into dry mining and wet mining. Dry processing techniques such as air cyclones, magnetic separation, ore sorting is applied to mineral sand industry due to the limitations like dust generation, lower efficiency in recovery (Napier-Munn and Morrison 2003). Due to advantages like ability in using the chemical properties of minerals while separation, high recover efficiency wet ore processing in mining with common techniques such as floatation, leaching, gravity separation, solvent extraction and electrowinning were used. The water inputs are quantified by the amount of processing material and the solid density requirement for individual processing. Many studies have been reported on copper (Northey et al. 2013), bauxite, iron, nickel, zinc, uranium and coal (Mudd 2010), platinum group metals (Glaister and Mudd 2010) with water consumption but the details such as sources of water, quality of water are not mentioned. The total water footprint during the process is not discussed in detail instead the blue water footprint alone for a platinum mine which is located in south Africa is discussed by Ranchod et al. (2015) by accounting evaporated water, water utilized by the product, water which does not return to the same catchment area. As for as evaporation is concerned, the evaporation during water storage, processing, collection as well as disposal are included. The blue water footprint estimation are as follows,

$$WF_{proc,blue} = Blue\ water\ evaporation + Blue\ water\ incorporation + Blue\ water\ lost\ return\ flow.$$

$$(10)$$

The water footprint for copper production located in El Teniente mine site (Olivares et al. 2012) and in Northern Chile (Pena and Huijbregts 2014) are also done.

4.4   Water Footprint in Manufacturing Industries

4.4.1   Water Footprint of Textile Industries

Textile industries account for the large consumption of water footprint all over the world. For example for 1 kg production of cotton fabric, one needs 10,000 L of water approximately and for the production of cotton t-shirt one needs 2,500 L of water in approximation (Freitas et al. 2017). Water is considered to be key resource in textile industries. The production of cloth depends upon three types of fibre, Polyester, Viscose and Cotton. Polyester and viscose which are manmade and cotton which is natural fibre. Polyester is a synthetic fibre which is used worldwide. The raw material for polyester may be crude oil or gas. Depending upon the process utilized for production, water consumption and wastewater are varied. Polyester are prepared using two methods, one is filament form and another one is staple form. In filament form, the minimum and maximum amount of green water utilized is 50 (m^3/ton and 52 m^3/ton) and the grey water obtained is 50,640 (m^3/ton) minimally and 71,033 (m^3/ton) in maximum. In stable form, the blue water consumed is 31 (m^3/ton) on average and 71,377 (m^3/ton) utilized maximum. On the whole in polyester production, blue water footprint are seen in fibre manufacturing process and the grey water footprint are seen in all production process, where oil explo- ration and refinery phases contributes in large amount. The raw material used for viscose production is cellulose which includes wood, cotton and bamboo. Trees are harvested for Viscose manufacturing. So, green water is utilized here. In viscose stable fibre production and for viscous filament yarn continuous as well as batch washing process, averagely 44 (m^3/ton) of green water are utilized separately. The blue water consumption in staple form is 156 (m^3/tone) in average a filament yarn production 370 (m^3/ton) were utilized. The grey water utilization in staple form is approximately 638 (m^3/ton) and in filament yarn with continuous washing, gen- erates 30,489 (m^3/ton) in maximum and in batch washing 3,489 (m^3/ton) in max- imum. For cotton manufacturing, the green and blue water footprint only considered and it varies with the geographical area. From the transformation of cloth from the raw materials it needs several litres of water. Let us see example of one apparel that is denim, which has become universal material. According to Luiken et al. (2015), the total water consumption is around 11,000 L per pair of jeans and the production of pairs of jean is 3.5 billion pairs and Gracia (2015) estimated that global jean production is estimated to be 5000 million units and the average water required for one pair of jean is 70 L. The water footprint of denim product refers to the total amount of water consumed in manufacturing of denim product. In a case study done by Wang et al. (2013) on textile industries, China, bottom-up approach was followed for calculating direct blue water footprint and direct grey water footprint. The $WF_{dir,blue}$ (Gm^3/year) can be calculated as follows,

$$WF_{dir,blue} = \sum WF_{dir,blue}[P] = \sum \begin{pmatrix} Direct\ Blue\ water\ evaporation\ \pm \\ Direct\ Blue\ water\ incorporation\ \pm \\ Direct\ lost\ rturn\ flow \end{pmatrix} [p]$$

(11)

$WF_{dir,blue}$ [p] represents $EF_{dir,blue}$ of industrial process p. After industrial process, the generated water from the process contain chemical oxygen demand, nutrients such as nitrogen and phosphorous, total suspended solids and salts which are of high concentrations and this process water accounts for grey water footprint. The calculation of $WF_{dir,grey}$ is not standardized and it needs further clarification. The degree of direct water pollutants are categorized into original $WF_{dir,grey}$ and residuary $WF_{res,grey}$ and the $WF_{dir,grey}$ and $WF_{res,grey}$ are as follows,

$$WF^{ori}_{dir,grey} = \sum WF^{ori}_{dir,grey}[p] = \sum \left(\max \left(\frac{L^k_{dir,ori}}{c^k_{max} - c^k_{nat}} \right) \right)[p] \quad (12)$$

$$WF^{res}_{dir,grey} = \sum WF^{res}_{dir,grey}[p] = \sum \left(\max \left(\frac{L^k_{dir,res}}{c^k_{max} - c^k_{nat}} \right) \right)[p] \quad (13)$$

$WF^{ori}_{dir,grey}$ Amount of fresh water of load of pollutants (Gm^3/year)

$WF^{res}_{dir,grey}$ Required amount of fresh water take in by the residuary pollutants (Gm^3/year)

$L^k_{dir,ori}$ Original load of pollutants generated directly in industrial process (t/year)

$L^k_{dir,res}$ Residuary pollutant load k after treatment (t/year)

Additionally, WF intensity have been introduced in this research work to analyses the water uses in textile industries in China.

$$WFI = \frac{WF}{TIOV} \quad (14)$$

where WFI measured in m^3/thousand US dollars and TIOV is the total industrial output value of the textile industry (Thaler et al. 2012; Vanham 2013).

c^k_{max} Maximum acceptable pollutant limit (mg/l)

c^k_{nat} Natural concentration of pollutant k in receiving water body (mg/l).

4.4.2 Water Footprint in Paper Industry

The estimated water footprint for paper is found to be 300–2600 m^3/ton (approximately 2–13 L of water for A4 sheet) (Van Oel and Hoekstra 2010). The green and blue water footprint of paper is determined by the rainwater which is evaporated during plant and tree growth. In water footprint in forestry and in industrial stage are calculated by Van Oel and Hoekstra (2011) and it is estimated as follows,

$$WF[p] = WF_{forestry}[p] + WF_{industry}[p] \quad (15)$$

The water footprint in forestry stage of paper product is estimated as,

$$WF_{forestry}[p] = \left(\frac{ET_a + (Y_{wood} + f_{water})}{Y_{wood}}\right) \cdot f_{paper} \times f_{value} \times \left(1 - f_{recycling}\right) \quad (16)$$

ET_a Evapotranspiration from forest or woodland (m^3/ha/year)
Y_{wood} Wood yield from woodland (m^3/ha/year)
f_{water} Volumetric fraction of water from harvested wood (m^3/m^3)
f_{paper} Wood-to-paper conversion factor (m^3/ton)
f_{value} The total fraction value of the forest associated with paper production
$f_{recycling}$ Pulp fraction derived from recycled paper.

$$f_{paper} = \frac{1}{f_p \times \rho} \quad (17)$$

ρ is the density of harvested wood (ton/m^3).

The water footprint for industrial stage in paper product is estimated as follows:

$$WF_{industry}[p] = E + R + P \quad (18)$$

E   Evaporation in the production process (m^3/ton)
R   Water in solid residuals (m^3/ton)
P   Water in products (m^3/ton).

In another research done by Rep (2011) on UPM paper mill located in Kymmene, accounted for water footprint. The green water footprint as suggested by Van Oel and Hoekstra (2010) was simplified by not-accounting the irrigation water and the fertilizers involved as these two makes small amount of contribution to the water footprint it has been excluded from the study. The green water footprint is consumed majorly in forestry and the estimation are as follows,

$$WF_{forestry}[P] \frac{ET_a + (Y_{wood} \times f_{water})}{Y_{wood}} \quad (19)$$

The fresh water utilized in paper industry from river, lakes and ground water comes under the category of blue water footprint. The assessment includes, (1) the evaporated water of drying section of paper and pulp production, (2) Stored water in paper and pulp product, (3) Water in the effluent sludge, (4) Losses which includes water leakage and non-retainment of water to the same catchment area.

The blue water footprint is calculated as,

$$WF_{proc,blue} = Blue\,water\,evaporation + Blue\,water\,incorporation \\ + Lost\,return\,flow\,[volume/time] \quad (20)$$

The fresh water needed for dilution of loaded pollutants before reaching to water bodies and the quantity of fresh water is decided by the pollutant concentrations and the standards of each pollutant, the grey water footprint are estimated as follows;

$$WF_{proc,grey} = \frac{L}{C_{max} - C_{nat}} = \frac{Effl \times C_{effl} - Abstr \times C_{act}}{C_{max} - C_{nat}} \tag{21}$$

L   Pollution load (mass/time)
Effl   Effluent volume noted in terms of time
Abstr   Abstraction water volume (Volume in terms of time)
C_{act} Actual concentration of water intake (mass/volume).

4.4.3 Water Footprint in Food and Beverages

Due to water scarcity and population increase, food and beverages industry are under high pressure. It is predicted that, between 2010 and 2050, people will be increased from 6.99 to 9 billion and 2.0 billion will be additional to feed. Food and beverages require higher water footprints. For the production of 1 kg of beef requires 15,500 L of water, 1 kg of chocolate requires 24,000 L of water, 1 kg of cheese requires 5,000 L of water (Arjen and Hoekstra 2008). Soft drinks, bottled water, wine and spirits consumes large amount of water footprint. In manufacturing process, water is consumed in production phase, refrigeration and in steam generation and finally in cleaning and maintaining phase. In final products, water is used as integral component. The food and beverages manufacturers which includes Nestle SA, Unilever Group, The Coca-Cola Company, Danone Group and Kraft Foods Inc. use 600.00 billion litres of water per year as per annual analysis done on 2008 which is equal to the water demand of people in entire earth. Some of the studies done by Coca-cola (Coca-Cola Europe 2011), Dole (Sikirica 2011), Unilever (Jefferies et al. 2012), Mars (Ridoutt et al. 2009). Generally, beverage industry estimates water quantity by water use ratio (WUR), which is total water usage divided by bottling production and finally expressed in terms of litre of water utilized by the litre of beverage produced. Food and beverages industries concentrated on water footprint reduction by practicing water recycling. Water footprint in dairy products apart from manufacturing are discussed in detail. The animal product starts with feed crop cultivation and ends with the consumer. In each step there is direct as well as indirect water footprint. The largest water footprint consumption is seen in the first step: growing the feed. The green and blue water footprint together in the harvested filed is equal to the evapotranspiration from the crop (m³/ha) divided by the crop yield (tons/ha). The grey water footprint of the crop is calculated using load of pollutants leaches to the groundwater (kg/ha) divided by standards of chemicals in quality of water (g/L) and the crop yield (ton/ha). The volume of water consumed for drinking and the feed consumed during its lifetime gives the water footprint throughout its life time. Around 98% of water footprint are used for feed (Mekonnen and Hoekstra 2010). Gerbens-Leenes et al. (2011) studied about the factors determining the water footprint of animal products and concluded with two factors. The first factor is feed conversion efficiency and the second factor is composition of feed taken up by the animals. The former describes about the volume of feed to produce the meat, eggs or milk and the later favor on grazing system. In a research done by Zareena (2016) Pondicherry Coop. Milk Supply Society, India was taken in this study. The daily production of milk was found to be 1.2 lakh L/day and depended on ground water for the production. The relationship between the productivity of water in production of milk and the cattle input and outputs are expressed as:

$$\sigma_{diary,j} = Q_{MP}/\Delta_{water} \tag{22}$$

Q_{MP} average daily milk yield over life cycle of livestock

$$\Delta_{water} = \omega_{df}/Q_{df} + \omega_{gf}/Q_{gf} + \Delta_{drink} \tag{23}$$

where,

Q_{df}, Q_{gf} average weights of green and dry fodder (kg/day)
Ω_{df}, ω_{gf} usage of water in dry and green fodder (L/kg/day).

Singh and Kumar (2004) showed average milk yield of two different types of cow namely, indigenous cow and crossbred cow. According to the study the green fodder and dry fodder needed for two types of cow are 12.92 (kg/day), 14.41 (kg/day) and the dry fodder quantity is 5.07 (kg/day) and 4.33 (kg/day). The average milk yield for indigenous cow and crossbred cow is 2.98 (L/day) and 4.46 (L/day).

5   WF Benchmarks Establishment and Reduction Goals

For the products like food and beverages, textile, agriculture water is utilized in larger amount and WF benchmarks are to be established for minimal water use. WF benchmarks provides the details of maximum consumptive use of water at each step of the product production stage with best available techniques. WF benchmarks provides reference for the government for allocation of WF permits to the industries. Government and business associations together should contribute some effort on establishing benchmarks and in framing the laws and legislations. By water recycling, reducing evaporation losses, utilizing the used chemicals in water flows, industries can reduce the WF. The WF gives the details of total amount of water consumption and the polluted quantity, WF has to be reduced to a considerable amount for sustainable development. Companies should set goals and targets regarding reduction of water footprint. In cases like agriculture and mining, achievement of zero WF is impossible but in the reduction of water consumption and pollution in water can be achieved (Brauman et al. 2013; Hoekstra 2017).

6 Challenges in WF Analysis

Sustainable allocation of water resources, water footprint is considered to be an important tool. ISO 14044 was framed on 2014 which provides standards, principles, requirements and guidelines of water footprint for processes and organizations and it also provides analyzing steps for spillage water as well as with consideration of local environmental conditions. By knowing the blue water nearby the processes and organization the blue water footprint can be noted and the estimation of rainwater intake and the intake of trees and plants are very critical to estimate. Till now, the approximation of inflows is maximum done in all research paper and regarding grey water footprint which is an indication of pollution on water quality, the parameter estimation in grey WFs is found to be tedious factor. In agriculture sector, lack of data in estimation of pollutant load in runoff is a major problem and also lack of data from the type of receiving water bodies is another tedious problem. Especially in grey water footprint, insufficient data regarding time and specific site location are seen. Apart from direct water footprint many industries often use indirect blue water which is higher than direct consumption. Depending upon the location and the operations used WFs per unit of product across the world. In order to achieve zero WF, balancing between water reduction and carbon footprint is necessary. In case of agriculture and mining, achievement of zero WF is impossible but reduction of water usage per unit of production has to be done (Brauman et al. 2013; Hoekstra 2015; Andreea and Teodosiu 2011).

7 Conclusion

The concept of water footprints (WF) gives an opportunity to link the production goods with the water resources. It gives an idea about the water consumption pattern and the dimensions in global aspect for good governance. At present, the demand of water footprints benchmarks for intensive goods, water pricing reflecting water scarcity, water footprint ceilings for each river basin are to be rectified for the sustainable water resources all over the world. This can be achieved by efficient usage of water by limiting the global water resources all over the communities and nations and the reduction of water by water footprint assessment is found to be an efficient and sustainable results as it gives clear cut idea about the utilization of water throughout the process.

References

Andreea, E. S., & Teodosiu, C. (2011). Grey water footprint assessment and challenges for its implementation. *Environmental Engineering and Management, 10*(3), 333–340.

Bhat, T. A. (2014). An analysis of demand and supply of water in India. *Journal of Environmental and Earth Science, 4*(11), 2224–3216.

Brauman, K. A., Siebert, S., & Foley, J. A. (2013). Improvements in crop water productivity increase water sustainability and food security: A global analysis. *Environmental Research Letters, 8*(2), 024–030(7).

Europe, Coca-Cola. (2011). *Water footprint sustainability assessment: Towards sustainable sugar sourcing in Europe*. Brussels: Belgium.

Dong, H. J., Geng, Y., Sarkis, J., Fujita, T., Okadera, T., & Xue, B. (2013). Regional water footprint evaluation in China: A case of Liaoning. *Science of the Total Environment, 442,* 215–224.

Ercin, A. E., & Hoekstra, A. Y. (2012). Carbon and water footprints: Concepts, methodologies and policy responses. United Nations World Water Assessment Programme, UNESCO; 2012. ISBN: 978-92-3-001095-9.

Freitas, A., Zhang, G., & Mathews, R. (2017). Water footprint assessment of polyster and viscose, C&A Foundation (2017).

Garcia, B. (2015). Reduced water washing of denim garments. In *Denim: Manufacture, finishing and applications* (pp. 405–423). Woodhead Publications.

Gerbens-Leenes, P. W., Mekonnen, M. M., & Hoekstra, A. Y. (2011). A comparative study on the water footprint of poultry, pork and beef in different countries and production systems. Value of Water Research Report Series No. 55. UNESCO-IHE, Delft, The Netherlands.

Glaister, B. J., & Mudd, G. M. (2010). The environmental costs of platinum-PGM mining and sustainability: Is the glass half-full or half-empty? *Minerals Engineering, 23,* 438–450.

Hoekstra, A. Y., & Hung, P. Q. (2002). Virtual water trade. A quantification of virtual waterflows between nations in relation to international trade. *International Expert Meeting on Virtual Water Trade* (Vol. 12, No. 11, pp. 1–244).

Hoekstra, A. Y. (2017). Water footprint assessment in supply chains. In: Y. Bouchery, C. Corbett, J. Fransoo, & T. Tan (Eds.), *Sustainable supply chains*. Springer Series in Supply Chain Management (Vol. 4, pp. 65–85). Cham: Springer.

Hoekstra, A. Y. (2008). *The water footprint of food*. The Netherlands: Twente Water Centre, University of Twente.

Hoekstra, A. Y. (2015). The water footprint of industry. In J. J. Klemes (Ed.), *Assessing and measuring environmental impact and sustainability* (pp. 221–254). USA: Waltham.

Hoekstra, A. Y. (2016). A critique on the water-scarcity weighted water footprint in LCA. *Ecological Indicators, 66,* 564–573.

Hoekstra, A. Y., Chapagain, A. K., Aldaya, M. M., & Mekonnen, M. M. (2011a). *The water footprint assessment manual*. London-Washington, DC: Earthscan.

Hoekstra, A. Y., Chapagain, A. K., Aldaya, M. M., & Mekonnen, M. M. (2011b). *The water footprint assessment manual*. Retrieved February, 2011, from http://doi.org/978-1-84971-279-8.

IEA (International Energy Agency). (2011). *The IEA Model of Short-Term Energy Security (MOSES) Primary Energy Sources and Secondary Fuels Working Paper*. Paris: OECD/IEA.

International organization for Standardization (ISO). (2014). https://www.iso.org/standard/43263.html.

Jefferies, D., Munoz, I., Hodges, J., King, V. J., Aldaya, M., Ercin, A. E., et al. (2012). Water footprint and life cycle assessment as approaches to assess potential impacts of products on water consumption: Key learning points from pilot studies on tea and margarine. *Journal of Cleaner Production, 33,* 155–166.

Lovarelli, D., Bacenetti, J., & Fiala, M. (2016). Water footprint of crop productions: A review. *Science of the Total Environment, 548–549,* 236–251.

Luiken, A., & Bouwhuis, G. (2015). Recovery and recycling of denim waste. In *Denim: Manufacture, finishing and applications* (pp. 527–540). Woodhead Publications.

Martínez-Paz, J. M., Pellicer-Martínez, F., & Colino, J. (2014). A probabilistic approach for the socioeconomic assessment of urban river rehabilitation projects. *Land Use Policy, 36,* 468–477.

McKinsey. (2009). *Charting our water future: economic frameworks to inform decision-making.* Munich: 2030 Water Resource Group, McKinsey Company.

Mekonnen, M. M., & Hoekstra, A. Y. (2010). The green, blue and grey water footprint of farm animals and animal products. Value of Water Research Report Series No. 48. Delft, The Netherlands: UNESCO-IHE.

Mekonnen, M. M., & Hoekstra, A. Y. (2012). A global assessment of the water footprint of farm animal products. *Ecosystems, 15*(3), 401–415.

Mudd, G. M. (2010). The environmental sustainability of mining in Australia: Key production trends and their environmental implications for the future. Research report No. RR5, Department of Civil Engineering, Monash University and Mineral Policy Institute (Revised April 2009).

Napier-Munn, T. J., & Morrison, R. D. (2003). The potential for the dry processing of ores. In *Proceedings of Water in Mining 2003* (pp. 247–250), Brisbane, QLD, 13–15 Oct.

Northey, S., Haque, N., & Mudd, G. (2013). Using sustainability reporting to assess the environmental footprint of copper mining. *Journal of Cleaner Production, 40,* 118–128.

Oel, P. R. V., & Hoekstra, A. R. (2010). The green and blue water footprint of paper products: Methodological considerations and quantification. Value of Water Research Report Series No. 46. Enschede, The Netherlands: ITC, University of Twente.

Olivares, M., Toledo, M., Acuna, A., & Garces, M. (2012). Preliminary estimate of the water footprint of copper concentrate production in central Chile. In *Proceedings of the Third International Congress on Water Management in the Mining Industry*, Santiago, Chile, 6–8 June 2012, Gecamin (pp. 390–400).

Pellicer-Martínez, F., & Martínez-Paz, J. M. (2016). Grey Water footprint assessment at the river basin level: Accounting method and case study in the Segura River Basin, Spain. *Ecological Indicators, 60,* 1173–1183.

Pena, C. A., & Huijbregts, M. A. J. (2014). The blue water footprint of primary copper production in Northern Chile. *Journal of Industrial Ecology, 18*(1), 49–58.

Pophare, A. M., Lamsoge, B. R., Katpatal, Y. B., & Nawale, V. P. (2014). Impact of over-exploitation on subwater quality: A case study from WR-2 Watershed, India. *Journal of Earth System Science, 123,* 1541–1566.

Ranchod, N., Sheridan, C. M., Pint, N., Slatter, K., & Harding, K. G. (2015). Assessing the blue-water footprint of an opencast platinum mine in South Africa. *Water SA, 41*(2), 287–293.

Rep, J. (2011). From forest to paper, the story of our water footprint, A case study for the UPM Nardland Papier mill, Kymmene, 2011.

Ridoutt, B. G., Eady, S. J., Sellahewa, J., Simons, L., & Bektash, R. (2009). Water footprinting at the product brand level: Case study and future challenges. *Journal of Cleaner Production, 17,* 1228–1235.

Rodriques, D. B., Gupta, H. V., & Mendiondo, E. M. (2014). A blue/green water based accounting framework for assessment of water security. *Water Resources Research, 50,* 7187–7205.

Morera, S., Corominas, L., Poch, M., Alday, M. M., & Comas, J. (2016). Water footprint assessment in wastewater treatment plants. *Journal of Cleaner Production, 112*(20), 4741–4748.

Sandoval-Solis, S., McKinney, D., & Loucks, D. (2011). Sustainability index for water resources planning and management. *Journal of Water Resources Planning and Management, 137,* 381–390.

Schneider, C. (2013). Three shades of water increasing water security with blue. *Green, and Gray Water.* http://dx.doi.org/10.2134/csa2013-58-10-1.

Sikirica, N. (2011). *Water footprint assessment bananas and pineapples.* Driebergen, The Netherlands: Dole Food Company, Soil & More International.

Singh, O. P., & Kumar, M. D. (2004, July). Impact of dairy farming on agricultural water productivity and irrigation water user. http://publications.iwmi.org/pdf/H042638.pdf.

Srinivasan, V., Seto, K. C., Emerson, R., & Gorelick, S. M. (2013). The impact of urbanization on water vulnerability: A coupled human-environment system approach for Chennai, India. Global Environment system approach for Chennai, India. *Global Environmental Change, 23*(1), 229–239.

Strauss, L., & Co. Water (online). (2016). http://levistrauss.com/sustainability/planet/.

Thaler, S., Zessner, M., Bertran De Lis, F., Kreuzinger, N., & Fehringer, R. (2012). Considerations on methodological challenges for water footprint calculations. *Water Science and Technology, 65*(7), 1258–1264.

United Nations. (2010). United Nations Resolution 64/292 The Human Right to Water and Sanitation. www.un.org.

Vanham, D. (2013). The water footprint of Austria for different diets. *Water Science and Technology, 67*(4), 824–830.

Vanham, D., & Bidoglio, G. (2013). A review on the indicator water footprint for the EU28. *Ecological Indicators, 26*(2013), 61–75.

Vorosmarty, C. J., Mclntyre, P., Gessner, M. O., Dudgeon, D., Prusevich, A., Green, P., et al. (2010). Global threats to human water security and river biodiversity. *Nature, 467*, 555–561.

Wang, L., Ding, X., & Wu, X. (2013). Blue and grey water footprint of textile industry in China. *Water Science and Technology, 68*(11), 2485–2491.

Water Footprint (online). (2016). http://waterfootprint.org.

Water Footprint and Its Growing Importance (online). (2017). https://www.thebalance.com/water-footprint-and-its-growing-importance-2878071.

Water Footprint Concept (online). (2015, Updated 2018). https://www.gktoday.in/academy/article/water-footprint-concept-of-blue-water-greenwater-grey-water.

WWAP, U. (2012). The United Nations World Water Development 42 Report 4: Managing Water under Uncertainty and Risk.

WWDR. (2015). The United Nations World Water Development Report 2015, Published by the United Nations Educational, Scientific and Cultural Organization, 7, place de Fontenoy, 75352 Paris 07 SP, France.

Corporate Water Footprint Accounting of Select Thermal Power Plants in India

Debrupa Chakraborty

Abstract Estimating water footprint for a production unit is knowledge intensive and appropriate data management task. It is important to know the total freshwater use by any consuming unit to be able to manage the resource better. Water footprint estimates provide one such single monitorable performance index. This chapter follows the component method to estimate water footprint (WF) for estimation of water footprint for two water intensive thermal power plants in India-one (old plant) in Dadri, Uttar Pradesh (U.P.) and the other (a new plant) in West Bengal. The total footprint of both the power plants constitutes of blue WF (operational and over-head)—that of the Dadri unit it is estimated to be 102×10^5 m^3 and of the power plant in West Bengal to be 16×10^6 m^3 for the year 2015–2016. However it is the make-up water in the production process (of 1% in case of Dadri unit and 0.80% for the unit in West Bengal) and evaporation loss [of 2% and 1.5% from Cooling Tower (CT) respectively in case of Dadri unit and unit in West Bengal] that are to be replenished in the form of freshwater in the following year. Finally based on the results and analysis a few recommendations have been made adopting which the units can strive to achieve zero water footprint milestones.

Keywords Water footprint · Thermal power generation units · Operational and overhead water · Footprint · Blue water footprint · India

1 Introduction

Water is becoming increasingly scarce all over the world. "Water" mainly refers to freshwater. Global primary energy demand is projected to increase by just over 50% between now and 2030 (FAO 2012). Freshwater withdrawals are predicted to increase by 50% by 2025 in developing countries and 18% in developed countries

D. Chakraborty (✉)
Department of Commerce, Netaji Nagar College, 170/436, N.S.C.Bose Road, Kolkata 700092, India
e-mail: chakraborty_debrupa@yahoo.com

© Springer Nature Singapore Pte Ltd. 2019   21
S. S. Muthu (ed.), *Environmental Water Footprints*,
Environmental Footprints and Eco-design of Products and Processes,
https://doi.org/10.1007/978-981-13-2454-3_2

(FAO 2012). Thus considering interlinked issues such as water, energy, climate change, and sustainable development and ecosystem services together, managing use of freshwater has become the need of the hour. Industrial, agriculture and domestic water and energy uses create an adverse impact on ecosystems. Efficient use of water resources can thereby help companies to gain an edge over their competitors and build a distinctive reputation in the market. By 2030 demand for water will exceed supply by 40% and half of the world's population will live in water scarce region (FAO 2012). The manifold increase in demand is driven by industry, population growth, change in food habits and ever increasing energy demands. Industries namely beverage sector, power generation, mining, pulp and paper sector—all heavily dependent on water are thus exposed to water scarcity directly. This is all more so as water is essential to carry out not only production (direct) activities but also for carrying out indirect activities like cooling, heating, transport, cleaning and allied activities (FAO 2012).

The Water Footprint (WF) of a product or company is equal to the sum of the WF of the products of the producer or that a company produces. Water consumption is not only in form of freshwater appropriation, water pollution is another form that should be accounted for. The major goal of calculating WF is to find out how one can reduce humanity's WF so as to achieve sustainability. The Water Footprint can be used to determine freshwater consumption and assess the environmental impact of wastewater created by industrial activities. WF not only considers the WF of enterprise itself but also takes the WF of external supply chain into account. Total WF relates to operational and supply chain of an enterprise. Generally WF of relevant processes in a production system is considered. This is because WF of some inputs in the up-stream of production process is difficult to calculate because of data unavailability. Thus, in industrial WF assessment only operational WF is taken into consideration. However if supply chain is consuming large volume of water (as in beverage sector) and unsustainably then it has an impact on the business water footprint as a whole. It is necessary to find out the areas where actions on water related issues along the supply chain are to be taken and also dialogues to be initiated with the relevant stakeholders. However communicating water use statistics to the general customers or public in general remains a challenge. This is more of so as water has a local impact (compared to carbon which has a global environmental impact) making the communication of water use and WF very context specific.

Current official statistics (Annual Survey of Industries-Factory Sector, Various Volumes) do not provide an opportunity to understand total water use by an industrial production unit. There are a number of methods suggested in the literature to understand natural resource use to arrive at accounting process and efforts are on towards that in India as well (Syamroy 2011). One method is water footprint estimate (Chakraborty 2012) which takes users perspective and considers direct and indirect use of freshwater for carrying out business activities. "Corporate water footprint" or "organizational footprint" is a measure of the volume of freshwater used at the place where the actual production and water use take place (Hoekstra and Chapagain 2007, 2008). The water footprint is normally expressed as green, blue and grey water footprints. The green water footprint refers to the consumption of rainwater stored in the soil as soil moisture. The blue water footprint refers to the evaporated surface and ground water. The grey water footprint refers to volume of freshwater required to assimilate the load of pollutants. Water Footprint Analysis (WFA) is an upcoming relevant concept used for the analysis of freshwater use, scarcity, and pollution in relation to consumption, production, and trade (Hoekstra 2017; Zhang et al. 2017). Water footprint can be developed for a variety of activities. It can be on an individual, family, village, city, province, state or nation (WBCSD 2006; Ma et al. 2006; Hoekstra and Chapagain 2007b) or producers e.g. a public organization, private enterprise or economic sector, for a specific activity, goods or services (Chapagain et al. 2006; Hoekstra and Chapagain 2007a), paper and paper consumption in Netherlands (Oel and Hoekstra 2012), agricultural crop (tomato) production at 24 farms in the Pinios river basin in Greece (Evangelou et al. 2016). Many companies have addressed the issues of water footprint and formulated proactive management strategies (Gerbens–Leenes et al. 2003). Since freshwater scarcity is considered a major risk to the global economy in terms of potential impact (WEF 2017) sustainable management of fresh water resources is a prerequisite for development. Business water accounting is increasingly becoming an integral part of sustainable corporate performance accounting so as to access water-use efficiency of production as a way to mitigate water scarcity. The study by Guzmán et al. (2017) explicitly points at water-use efficiency (WF per unit of product). In India, water used in the consumption of agricultural goods has been analysed (Kampman 2007). Also the study of the inter- state virtual water flows in India has been conducted (Verma et al. 2009). Water account using UN method has also been tried (Syamroy 2011). WF has been quantified for a water intensive paper production unit in West Bengal, India (Chakraborty and Roy 2012). These are all still very preliminary standalone attempts. In developing countries like India, water footprint can be used as indicator for sustainable water management, especially for industries in the face of competing demand for water. Accounting for corporate water use through the application of water footprint concept can identify the business water related risks. This can also influence the business strategies and help towards formulating water policy relevant for business sector. There is dearth of studies on water footprint for Indian industries. We could not get any secondary source of information, which could provide us with dependable estimate of water footprint by industries. Thus primary data information collected from chosen study units. The case study units provides good representation of Indian thermal power generating plants and belongs to companies that are fore runner so far as modernization and sustainable development goals are concerned. This chapter demonstrates the ways where WF can be estimated using unit level primary data collected from the specific industrial units. Final choice of the study unit has been determined by willingness of the unit to cooperate in data sharing and time commitment for verbal communication.

Global water demand in terms of water withdrawal are estimated to increase by 55% by 2050, the reason behind which is identified to be growing demands from manufacturing (400%), thermal electricity generation (140%) and domestic use

(130%). Water plays a number of important functions varying from being used as a raw material in pharmaceutical industry to cleaning manufacturing facilities and for cooling in power generation stations. However focus of any industry is on its primary business not on energy and water efficiency. But these should be incorporated into its larger objectives in order to achieve cost control and satisfying corporate and social responsibilities (Walsh 2015).

In this backdrop, the objectives of this study are

 (i) To estimate the water footprint of water intensive industrial units in India. Two thermal power generating units—one old [situated in Uttar Pradesh (U.P.) in the Northern part of India] and one new [situated in West Bengal in the eastern part of India] has been considered as case study.

 (ii) To study the water management of the water consumption and pollution along the whole production chain of the industrial units so as to identify the critical WF components contributing towards water footprint. This has been done by comparing the WF of the concerned industries with that of an existing benchmark.

(iii) Benchmarking of environmental performance of the case study units by evaluating how these business activities can become more sustainable by analyzing response strategies and formulating reduction targets (if necessary) in quantitative and/or qualitative terms.

(iv) Finally, this study aims to provide a comprehensive scope of analysis of the water footprint resulting from the case study unit, the results from which can help towards decision making and identification of "hotspots" and sustainable options.

Challenges or Limitations of the Study:

 (i) Direct Water footprint: Calculation of blue water footprint made based on certain assumptions due to non-availability of detailed data in terms of consumption of freshwater.

 (ii) Indirect water footprint: Dearth of knowledge of supply chain because of unwillingness of suppliers to share data relating to water consumption and pollution for the case study power plants. Availability of indirect WF data could have helped in providing a more accurate estimation of WF of the case study units.

(iii) Response Strategy Formulation: Difficulties have been faced in making a quantitative analysis of WF reduction possibilities relating to response strategies.

The remainder of the chapter is organized in four parts. Section 2 encompasses a brief introduction and the history of case study units, Sect. 3 constitutes the materials and methodology and Sect. 4 provides discussion and analysis of results obtained from thermal power production units. Section 5 concludes the chapter with some recommendations of the study in brief.

2 Brief Introduction of the Thermal Power Generation Units

Electricity production contributes towards 37% of the total global emissions according to the IPCC (www.world.nuclear.org) and electricity demand is expected to increase by 43% over the next 20 years (WNA 2011). There are a number of different electrical generation methods, each one of which produces greenhouse gases (GHG) in varying quantities through construction, operation and decommissioning. Coal based generation power plants releases the majority of GHGs during operation whereas wind and nuclear power plants release the majority of emissions during construction and decommissioning. Indian power sector has recorded an impressive growth over the last few decades from 1,713 MW in 1950 to 245,258.54 MW as on 31.03.2014 out of which coal based thermal power plant occupies a lion's share of 145273.39 MW (CEA 2014). Out of the total capacity generation of 245,258.54 MW in 2014 sector wise distribution of installed capacity reveals that thermal power plant occupies a lion's share 168254.99 MW) followed by hydro (40531.41 MW) taking the second position and renewable (31692.14 MW) and nuclear (4780.00 MW) occupying the third and fourth positions (CEA 2014). Again, installed capacity is predominantly coal based and therefore, is a major source of carbon dioxide emissions in India. For producing power at a thermal power plant, water is one of the key input or component along with coal. The requirement of water for other plant consumptive uses is met from an alternative source or by installing desalination plant (CEA 2014). Water consumption in plants are governed by a number of factors such as quality of raw water, type of condenser cooling system, quality of coal, ash utilization, type of ash disposal system, waste water management aspects etc. Water use in thermal power plant is dominated by cooling which is met with treated recycled water. Thus, for plants with similar heat rates, the types of cooling system used in generation plant have greater effect on water consumption intensity than the type of fuel used. Water usage matters when water used by power sectors is freshwater as against irrigation or municipal water. If power plants use alternative water sources such as waste water or saline the impact can be mitigated.

Thermal Power Plant in Dadri, U.P.

The company to which this plant belongs is India's largest energy conglomerate established in 1975 to accelerate power development in India. It is one of the biggest power plants with an installed capacity of 51,635 MW. The Company has been operating its plant at high efficiency levels. The company as a whole envisages a generation mix by evaluating conventional and alternative sources of energy to sustain long run competiveness and mitigating fuel mix.

The Dadri unit situated in Uttar Pradesh (U.P.) in the Northern part of India is an old and unique power plant (installing a distinctive water saving SCALE-BAN™ Equipment) which has coal based thermal plant of 1820 MW [4 units × 205 MW + 2 units × 500 MW], gas based thermal plant of 817 and 5 MW solar plant totalling 2642 MW. The coal based thermal unit generating an annual output of

9600 MU (Million kWh) in 2015–2016 has been under consideration in this study (taking 365 working days). The functional unit is 9600 Million Unit (MU) 1MU = 10,00,000 kWh of generated electricity. The unit extracts freshwater from the nearby river body.

Thermal Power Plant in West Bengal

The case study thermal power plant situated in West Bengal in the eastern part of India is a new plant which started operating from 1999 with two generating units initially and later introduced three more units one in 2000 and the other two in 2007 and 2008 respectively. This plant covers a total area of 750 acres and built up area of 600 acres. It is one of the most reliable and prestigious coal-fired power plants in West Bengal and in India as well. Funded by the Over-seas Economic Co-Operation Fund (OECF) of Japan Govt.—sub-sequently constituted as Japan Bank for International Co-operation (JBIC)—this project is one of the first Fast Track projects to be successfully completed within scheduled time. In two stages the total generation capacity of the plant is (5 units × 210 MW) i.e. 1050 MW. The unit has a total generating capacity of 7300 Million Unit (MU) of electricity. However during the year 2015–16 (taking 365 working days) the power plant generated 6669 Million Unit (MU) of electricity. The functional unit is 6669 Million Unit (MU) 1 MU = 10,00,000 kWh of generated electricity. This plant is also another unique case study unit because the plant uses rain water from the two nearby dams to meet its water consumption partially. One of these two dams has no link to any river body whereas the other dam has link to a nearby river. As a result freshwater extracted from river is used in the production process and the water used in cooling tower is partially withdrawn from the rain water fed dams and the rest met from recycled water. The study units chosen for the present study is a good representation of Indian thermal power generating units and belongs to companies that are forerunner so far as modernization and sustainable development goals (rainwater harvesting being a unique feature in case of the unit in West Bengal) are concerned.

3 Materials and Methods

3.1 Water Footprint Assessment Methods, Tools and Standards

1. Calculation Method	(i) **Bottom-up Approach**: This is an Item-By-Item Approach (Hoekstra and Chapagain 2008). In this approach, the WF is calculated by multiplying all goods and services consumed by the people of a country by their respective water needs for those goods and services The component-based method is more suitable for the assessment of the footprint of an individual, business or sub-national community where import–export data are not available

(continued)

(continued)

	(ii) **Top Down Approach**: The Compound calculation or the method of top-down accounting in WFA, which is based on drawing national virtual water trade balances method (Hoekstra 2007). Here the WF of a nation is calculated as: Total use of domestic water resources + gross virtual water import − gross virtual water export The compound based method is suitable for sector, national and global studies
2. WF Standard	In WF accounting, there is only one standard: the Global Water Footprint Standard published by the Water Footprint Network (WFN) in 2009, revised in 2011 (Hoekstra et al. 2011) and 2014. The revised standard of 2014 standard (ISO 14046:2014) specifies principles, requirement and guidelines related to WF Assessment of products, processes and organization based on Life Cycle Assessment (LCA). The standard can be used for nation, province, municipalities and can help government to manage water resources and achieve sustainable development (Ercin and Hoekstra 2012)
3.Water Footprint Assessment Tools	The WF assessment tool is a free online web application published by WF Network in collaboration with University of Twente, Netherlands to assist the companies, government, NGOs, investors, researchers to calculate and map the WF. This in turn helps to identify actions to improve the sustainability and efficiency of water use (www.waterfootprint.org/tool/home/) However for calculating WF for agricultural crops and for managing climate risk in agriculture a tool—Agro Climate, developed by South East Climate Consortium (SECC). This consortium is a coalition of eight Universities in U.S. and is currently maintained by University of Florida. This tool helps to provide climate information to improve crop management decisions and reduce production risks associated with climate change (www.waterfootprint.org/tool/home/)
4. Sustainability of the Footprint	Additional information is required to assess sustainability of the WF. Per catchment area, freshwater availability and waste assimilation capacity need to be estimated, which form a WF cap for the catchment? For specific processes and products, WF benchmarks can be used (Ercin and Hoekstra 2012)

3.2 Methodology and Materials Used for Calculation of WF of Case Study Thermal Power Production Units

Water is used in almost all areas/facilities of thermal power stations in one way or other. A typical list of plant systems/applications requiring consumptive water is indicated as below:

  (i) Cooling water system for condenser and plant auxiliaries
 (ii) Ash handling system
(iii) Power cycle make up
 (iv) Equipment cooling system
  (v) Air conditioning and ventilation system
 (vi) Coal dust suppression system
(vii) Service water system
(viii) Gardening
 (ix) Evaporation from raw water reservoir.

For calculation of WF of case study thermal power plants the following types of water consumptions have been taken into consideration:

  (i) Processing—water in the boiler required for generating steam to move the turbine, cooling and cleaning of equipment, power cycle make up, coal dust suppression, ash handling due to blow down of relatively impure fluids from the water used in the process.
 (ii) Cooling Towers (CT)—for cooling the heated water to reduce evaporation and heat loss and CT system make up loss mainly due to evaporation loss in cooling towers.
(iii) Service water system for domestic purpose and gardening.

Method adopted for calculating the WF of thermal power generation units is the component-based method or bottom-up approach (Leenes and Hoekstra 2008). However with some modifications have been incorporated in this part. This method is used for the assessment of the water footprint of an individual, business or a single facility. This is found to be the most appropriate after we reviewed all other methods applied in various studies listed in the previous section.

   Let,

WF	Water Footprint
BWF_O	Operational Water Footprint
BWF_s	Suppy-Chain Water Footprint
$WF_{bus,oper,input}$	Operational WF for production inputs
$WF_{bus,oper,overhead}$	Operational WF for overheads
$WF_{bus,sup,input}$	Suppy-chain WF for production inputs
$WF_{bus,sup,overhead}$	Suppy-chain WF for overheads
BWF_{green}	Green WF
BWF_{blue}	Blue WF
BWF_{grey}	Grey WF.

Because the production units under study are generating thermal power, so green water footprint is not relevant so only blue and grey water footprints are estimated. WF is calculated by adding the Operational WF (direct *water use*) and Supply

Chain WF (indirect *water use*). Both Operational and Supply-Chain WF consist of two parts: the water footprint directly associated with the production of the product in the business unit and an overhead water footprint. The following relations explain the methods of estimation.

$$WF = BWF_O + BWF_s \tag{1}$$

$$WF_{bus,oper} = WF_{bus,oper,input} + WF_{bus,oper,overhead} \tag{2}$$

$$WF_{bus,sup} = WF_{bus,sup,input} + WF_{bus,sup,overhead} \tag{3}$$

Both in case of operational and supply-chain water footprint distinction is to be made between green, blue and grey water footprint by presenting the results with the help of the following formulae.

$$BWF_{bus,oper,input} = BWF_{o.green} + BWF_{o.blue} + BWF_{o.grey} \tag{4}$$

$$BWF_{bus,oper,overhead} = BWF_{o.green} + BWF_{o.blue} + BWF_{o.grey} \tag{5}$$

$$WF_{bus,sup,input} = BWF_{s.green} + BWF_{s.blue} + BWF_{s.grey} \tag{6}$$

$$WF_{bus,sup,overhead} = BWF_{s.green} + BWF_{s.blue} + BWF_{s.grey} \tag{7}$$

Finally, the total footprint of the business unit (BWF) is given by the sum of its operational (BWF_O) and supply-chain water footprint (BWF_S).

In this case study both production and overhead WF have blue WF components because green component will be of zero value. It is important to understand each of the components well to be able to compile relevant data. For WF calculation of thermal power plants, green water footprint is not relevant. Also grey water footprint is also not of much importance as thermal power plants hardly discharge any grey or waste water into the environment. This is because the fresh water extracted from the water bodies or from the ground is generally used in the process of generating heat. The heated water is then transferred to cooling towers and the cold water is recycled and used again in the power generation process. The effluent water generated in the dimineralising plant (DM) while generating heat is treated in the effluent treatment plant (ETP) and reused again in the power generation system, for irrigation purpose in the nearby areas and returned to the river thereby generating no grey water. So only blue water footprint is relevant in this case study. WF is calculated by adding the Operational WF (direct *water use*) and Supply Chain WF (indirect *water use*). Both Operational and Supply-Chain WF consist of two parts: the water footprint directly associated with the production of the product in the business unit and an overhead water footprint. *However only the blue component of operational and overhead WF is calculated in this study.* Supply-chain WF has not being taken into consideration due to non-availability of detailed data.

3.3   Operational WF

Operational WF consists of Operational WF directly associated with the production of the product (power generation in this case) in the business unit and an Overhead WF. Operational blue water footprints of the products [see Eq. 4] include the sources from which water is used by the power plant, water consumed during the power generation process and in the cooling tower. An overhead WF related to for example water consumed by employee's mainly drinking water, in kitchen, toilets, cleaning, gardening or washing working clothes (overhead blue WF) [refer to Eq. 5]. With this definition, working with the environmental department of the case study power plants detailed breakup of water use had to be collected.

The case study power plant in Dadri, U.P. generated 9600 MU (Million kWh) in 2015–2016, the year under consideration in this study. The coal based power plants have being considered for study excluding the gas and solar based power plants. Table 1 summarizes all the data that gathered from the Dadri Unit, U.P. for the year 2015–16.

The data given in Table 1 is represented in the form of a flow chart as depicted in the Fig. 1.

Table 2 summarizes all the data that collected from the power plant of West Bengal for the year 2015–16.

The data given in Table 2 is represented in the form of a flow chart as depicted in the Fig. 2.

Table 1 Data collected and used in calculating operational water footprint of the Thermal Power Production Unit, Dadri U.P. for the year 2015–16 (for 9600 Million KWh generation by 1820 MW Plant)

Area of water consumption	Freshwater consumption (m³/year)	(Treated wastewater) (m³/year)
1. Returned to agricultural land nearby river	–	–
2. Processing (in Boiler, condenser and deactorator)	93×10^5 m³	–
3. Cooling (Cooling tower)		249×10^5 m³^{3a}
4. Domestic (Kichen, drinking, gardening, road cleaning, sanitary floor cleaning)	19×10^5 m³	19×10^5 m³
Total	112×10^5 m³	$(-) 10 \times 10^5$ m³

Source Primary information collected from production unit under study through repeated visits and face to face interview with management
[a]Treated Recycled water is used in Cooling Tower

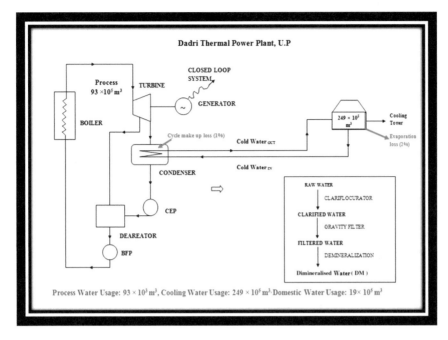

Fig. 1 Water Balance of Thermal Power Production Unit, Dadri U.P. for the year 2015–16

Table 2 Data collected and used in calculating operational water footprint of the Thermal Power Production Unit West Bengal for the year 2015–16 (for 6669 Million KWh generation by 1050 MW plant)

Area of water consumption	Freshwater consumption (m³/year)	(Harvested rainwater/ Treated wastewater used) (m³/year)
1. Processing (in Boiler, condenser and deactorator)	24×10^6 m³	–
2. Cooling (Cooling tower)		55×10^6 m³
3. Domestic (Kitchen, drinking, gardening, road cleaning, sanitary floor cleaning)	5×10^6 m³	
4. Harvested rainwater used in cooling tower		$(-) 25 \times 10^6$ m³
5. Treated sewage water returned to agricultural land		$(-) 13 \times 10^6$ m³
Total	29×10^6 m³	$(-) 13 \times 10^6$ m³

Source Primary information collected from production unit under study through repeated visits and face to face interview with management

Out of the total Cooling Tower (CT) water 55×10^6 m³, 25×10^6 m³ is from harvested rain water and rest 30×10^6 m³ and this is from recycled water

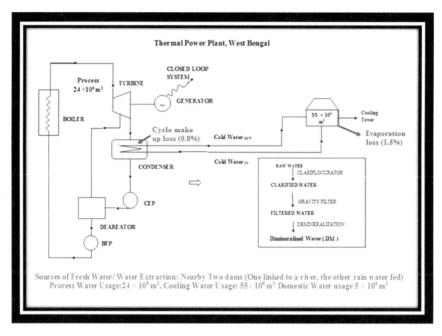

Fig. 2 Water Balance of Thermal Power Production Unit, West Bengal for the year 2015–16

4 Data Analysis and Findings

4.1 Water Consumption and Water Footprint Accounting of Thermal Power Plant, Dadri U.P.

Cooling is the main contributor in the total water consumption (72%) by the unit during the year 2015–2016. Treated recycled water is used for cooling. Water used in the production process contributes (26%). Process water includes cycle make up water or water used for demineralization (1%) in Demineralization (DM) Plant and Cooling water includes make up water or evaporation loss (2%) from Cooling Tower (CT) while overhead WF is negligible (2%) as depicted in Fig. 1. Closed loop cooling system is used for preventing evaporation loss and water usage. So in the next year cycle make up water of 1% and evaporation loss from CT (make up water of CT) of 2% is to be replenished in the form of freshwater.

Water footprint of Dadri, U.P. (m³/year) in 2015–2016

Direct Blue Water Footprint (Operational & Overhead WF)

Freshwater used in production process + Freshwater used for Domestic purpose − Treated water supplied for irrigation purpose or returned to the river=93 × 10⁵ m³ + 19 × 10⁵ m³ − 10 × 10⁵ m³ (as per data in Table 2)

Out of the total water used for domestic purpose (19×10^5 m^3), treated sewerage water (10×10^5 m^3) is used for irrigation and horticulture purpose in the nearby area.

To sum up, by returning to Eq. (4) we can provide the following numbers for the case study unit:

$$BWF_{bus,oper,input} = BWF_{o.blue}$$
$$= 93 \times 10^5 \text{ m}^3 \text{ [as } 249 \times 10^5 \text{ m}^3 \text{ recycled water is used in cooling towers]}$$

Using Eq. (5) the overhead operational WF can be shown as

$$BWF_{bus,oper,overhead} = BWF_{o.blue}$$
$$= 19 \times 10^5 \text{ m}^3 \text{ minus } 10 \times 10^5 \text{ m}^3$$
$$= 9 \times 10^5 \text{ m}^3$$

So, total Operational Water Footprint (inputs + overhead) of the unit (as per Eq. 2)

$$WF_{bus,oper} = WF_{bus,oper,inpus} + WF_{bus,oper,overhead}$$
$$= 93 \times 10^5 \text{m}^3 + 9 \times 10^5 \text{m}^3$$
$$= 102 \times 10^5 \text{m}^3$$

Operational Blue WF (93×10^5 m^3) constitutes 91.17% and overhead Blue WF (9×10^5 m^3) constitutes the rest 8.82% of the total WF for the year 2015–2016.

Table 3 presents the results of the total Water Footprint of the Thermal Power Production Unit, Dadri, U.P.

Total WF for 1820 MW plant generating 9600 million kWh of electricity is 102×10^5 m^3 for the year 2015–16. So water consumption per MWh of generation is 102×10^5 m^3/year divided by 365 days, 24 h and 1820 MW which equals to 2.17 m^3 i.e. **0.63 m^3/MWh**.

Table 3 Water Footprint of Thermal Power Production Unit, Dadri, U.P. for the year 2015–16

Information needed	Water Footprint (m^3/year) for 9600 Million KWh generation by 1820 MW plant			
	Green	Blue	Grey	Total
Operational Water Footprint	0	102×10^5	0	102×10^5

4.2   Water Consumption and Water Footprint Accounting of Thermal Power Plant in West Bengal

In case of power plant in West Bengal, it is cooling which is the main contributor to the total water consumption (65.47%) during 2015–2016. Water used in the production process contributes (28.57%) while overhead WF is negligible (5.95%). Process water includes cycle make up water or water used for demineralization (0.80%) and cooling water includes Cooling Tower (CT) make up water or evaporation loss (1.5%) as depicted in Fig. 2. Closed loop evaporative cooling system is used to reduce the total water usage. So in the next year cycle make up water of 0.80% and CT make up water of 1.5% is to be replenished in the form of freshwater. Treated recycled water and harvested rainwater is used for cooling system.

Water Footprint (m³/year) in West Bengal Power Plant During 2015–2016

Direct Blue Water Footprint (Operational & Overhead WF)

Freshwater used in production process + Freshwater used in Domestic purpose − treated sewage water is used for irrigation and horticulture purpose in the nearby area (as per data in Table 2).

$$= 24 \times 10^6 \, m^3 + 5 \times 10^6 \, m^3 - 13 \times 10^6 \, m^3$$

Out of the total water used for production, sewerage water ($13 \times 10^6 \, m^3$) is used for irrigation and horticulture purpose to the nearby area.

Also out of the total water used for cooling ($55 \times 10^6 \, m^3$) harvested rainwater has been used to the extent of $25 \times 10^6 \, m^3$ and the remaining $35 \times 10^6 \, m^3$ is from recycled water.

To sum up by returning to Eq. (4) we can provide the following numbers for the case study unit:

$$BWF_{bus,oper,input} = BWF_{o.}$$
$$= 24 \times 10^6 m^3 \, [\text{as treated recycled water and harvested rainwater of}$$
$$55 \times 10^6 \, m^3 \text{ is used in cooling tower}]$$

Using Eq. (5) the overhead operational BWF can be shown as

$$BWF_{bus,oper,overhead} = BWF_{o.blue}$$
$$= 5 \times 10^6 \, m^3$$

So, total Operational Water Footprint (inputs + overhead) of the unit (as per Eq. 2)

$$WF_{bus,oper} = WF_{bus,oper,inpus} + WF_{bus,oper,overhead} - \text{water returned for irrigation and}$$
$$\text{horticulture purpose to the nearby area}$$
$$= 24 \times 10^6 \, m^3 + 5 \times 10^6 \, m^3 - 13 \times 10^6 \, m^3$$
$$= 29 \times 10^6 \, m^3 - 13 \times 10^6 \, m^3$$
$$= 16 \times 10^6 \, m^3$$

Operational Blue WF ($11 \times 10^6 \, m^3$) constitutes 68.75% and overhead Blue WF ($5 \times 10^6 \, m^3$) constitutes the rest 31.25% of the total WF for the year 2015–2016.

Table 4 presents the results of the total Water Footprint of the Thermal Power Production Unit, West Bengal.

Total WF for 1050 MW plant generating 6669 million kWh of electricity is $16 \times 10^6 \, m^3$ for the year 2015–16. However water consumption per MWh of generation is calculated by taking on into consideration freshwater used *less* treated water returned to hydrological system for irrigation and horticulture purpose in the nearby area. Freshwater consumption amounting to $16 \times 10^6 \, m^3$/year divided by 365 days, 24 h and 1050 MW which equals to **1.73 m^3/MWh**.

Green and grey water footprint is nil in case of both the plants. For WF calculation of thermal power plants, green water footprint is not relevant. Also grey water footprint is also not of much importance as thermal power plants hardly discharge any grey or waste water into the environment. The effluent water generated in the demineralizing plant (DM) while generating heat is treated in the effluent treatment plant (ETP) and reused again in the power generation system and cooling towers, for irrigation purpose in the nearby areas and returned to the river thereby generating no grey water. The generated ash is used in cement industry, brick making, filling up of low lying areas etc. For both the thermal power plants the supply chain water footprint (Indirect WF) could not be estimated due to dearth of detailed data. This was a major limitation as estimating both the direct and indirect WF for any facility is essential for prioritizing the response strategies by the facility managers.

Table 4 Water Footprint of Thermal Power Production Unit, West Bengal for the year 2015–16

Information needed	Water Footprint (m^3/year) for 6669 million KWh generation by 1050 MW plant			
	Green	Blue	Grey	Total
Operational Water Footprint	0	$16 \times 10^6 \, m^3$	0	$16 \times 10^6 \, m^3$

4.3   Response Strategies of Thermal Power Plant in Dadri, U.P. and West Bengal

Water footprint accounting revealed a number response strategies that Dadri Unit have adopted. These include adoption of water conservation measures like using drift eliminator in cooling system, plant waste water is re-used for purposes like dust suspension system, cleaning, sewerage water is used for horticulture and irrigation in nearby areas. Water saving measure includes installation of SCALE-BAN™ Equipment which is a static, mechanical equipment having no limitation of hardness. SCALE-BAN™ has helped the plant in achieving reduced water consumption/MWh as per the guidelines issued by ministry of Ministry of Environment and Forest (MOEF). Water consumption in cooling tower is reduced by more than 25–30% and fresh water consumption has been brought down to a significant level.

Two types of ash are generated from thermal power plants—Ash pond and Dry ash. In case of ash pond ash water is recycled through Ash Water Re-circulating System (AWRS). The ash handling system of the plant forms a slurry (mixing ash with water) which is pumped out into ash pond. The ash in the pond solidifies and the clean water (containing minerals) is discharged into the nearby fields for irrigation and also used for recirculation of ash water. Fly ash or dry ash is sold for cement manufacturing and brick making Dadri's ash-handling approach is a trendsetter for South Asia. For the past 10 years, it has achieved almost 100% recycling of fly ash, mostly at an on-site brick manufacturing plant. In addition to bricks, more than 20 other products are produced at the plant. Fly ash is also supplied to cement manufacturers. Dadri was the first plant in Asia to use dry ash handling for bottom ash, and the landfills are carefully landscaped to prevent fugitive dust problems.

The response strategies of the power plant in West Bengal include wastewater treatment and recycling of treated wastewater and most eminent of all is rainwater harvesting carried out for the last 14–15 years. The harvested rainwater capacity of the plant is 25×10^6 m^3 which along with the water from nearby dam has been used in the cooling tower during 2015–2016. Out of the total ash generated in 2015–16, 40% is dry fly ash that has been sold off to different agencies and industries for manufacturing of environment friendly ash, block, tiles and road construction and thereby generating revenue all at once. On the other hand the balance 60% of the generated ash constitutes bottom ash (generated during the production process) which is disposed off to ash pond maintained for the said purpose. The ash handling system of the plant forms a slurry (mixing ash with water) which is pumped out into ash pond. The ash in the pond solidifies and the clean water is used in the system. This ash is transported to the nearby areas and used for filling up low lying areas and in road construction.

Another commendable response strategy of the two thermal power plants is that they have been able to meet the water consumption limit prescribed by the Ministry of Environment, Forest and Climate Change (MoEFCC) in its notification in official

Table 5 Water Consumption limit prescribed by the Ministry of Environment, Forest and Climate Change (MOEFCC) for Thermal Power Plants (in December, 2015), Government of India

Parameter	Standard prescribed	Standard achieved
Water	All plants shall install cooling towers (CT) and achieve specific water consumption maximum 3.5 m³/MWh within 2 years from the date of notification (December, 2015)	(i) Specific water consumption for 1820 MW plant at Dadri, UP have been 0.63 m³/MWh for the year 2015–2016 (ii) Whereas for the 1050 MW thermal power plant in West Bengal water consumption have been 1.73 m³/ MWh for the year 2015–2016

gazette in December, 2015 The prescribed standard and the standard achieved by the power plants are as follows (Table 5).

Both the plant have closed loop circulating system and cooling towers abiding the norms set by the Ministry of Environment, Forest and Climate Change.

5   Conclusions and Recommendations

5.1   Conclusions

Estimating water footprint for a production unit is knowledge intensive and appropriate data management task. It needs effort, coordination and capability to monitor the water use. This is all more so because efficient water use is an essential element in sustainable water use. Efficiency can be achieved by using less water per unit of production. The water use efficiency or productivity can measured by dividing total production by blue water footprint of an industry which means production per unit of water consumption. This chapter demonstrates the ways where WF can be estimated using unit level data collected from the specific water intensive thermal power plants.

The Dadri unit is a unique power plant of the case study company. Total WF for 1820 MW plant generating 9600 million kWh of electricity is 102×10^5 m³ for the year 2015–16. *Operational Blue WF* (93×10^5 m³) constitutes 91.17% and overhead Blue WF (9×10^5 m³/year) constitutes the rest 8.82% of the total WF for the year 2015–2016. Freshwater consumption per MWh of generation equals to *0.63 m³/MWh*. However it is the makeup water of 1% and evaporation loss from CT (2%) during 2015–16 is to be replenished in the form of freshwater. In other words 3% of total water consumption of 102×10^5 m³ is the actual water that is to be replenished with freshwater in the next year, as the total water consumed is actually re-circulated in the closed loop based system of power generation in these thermal power plants.

The plant in West Bengal on the other hand is also another unique case study unit because the plant uses harvested rain water from the two nearby dams to meet its cooling water consumption. As a result freshwater extracted from river is used in production process and the water used in cooling tower is withdrawn from the nearby dams and harvested rainwater is used along with treated effluent water. Total WF for 1050 MW plant generating 6669 million KWh of electricity is 16×10^6 m^3/year for the year 2015–16. *Operational Blue WF* (11×10^6 m^3) constitutes 68.75% and overhead Blue WF (5×10^6 m^3/year) constitutes the rest 31.25% of the total WF for the year 2015–2016. Water consumption per MWh of generation is *1.73 m^3/MWh* during 2015–2016. This is because recycled water, harvested rain water and water from the nearby dam is used in cooling tower. However, in the next year cycle make up water of 0.80% and CT make up water of 1.5% is to be replenished in the form of freshwater. Specific water consumption both the plants are well below the water consumption limit prescribed by the Ministry of Environment, Forest and Climate Change (MOEFCC) in its notification in Official Gazette in December, 2015. Both the plant have closed loop circulating system and cooling towers abiding the norms set by the Ministry of Environment, Forest and Climate Change.

On the other hand Dadri unit in-spite of generating more MW of electricity has a lower specific water consumption of 0.63 m^3/MWh as compared to the power plant in West Bengal. This is because Dadri Unit has installed unique water saving Equipment which has helped the plant in achieving reduced water consumption per MWh as per the guidelines issued by MOEFCC.

5.2   Recommendations

Water footprint accounting have led to a number response strategies that the case study units can apply in the future. These strategies include:

Thermal Power Plant at Dadri, U.P.

Recycling blow-down water, redesigning the Ash Water Re-circulating System (AWRS), cooling tower and condenser improvements, changing from wet to dry cooling systems for air conditioning cooling towers, and minimizing water use for ash management. The plant also have plans to improve sustainability of the watershed jointly with other power plants of the Company and suppliers.

Thermal Power Plant in West Bengal

Recycling of ash pond water of the wet ash slurry system for reuse should be increased; quantity of harvested rainwater used in the cooling system should also be enhanced. Water use for ash management to be minimized. More water saving advanced technologies should be introduced as the plant is dependent on the rain water and water from nearby dams for its supply of water to cooling towers. Also reduction in water requirement in the production process has to be reduced through adoption of water conservation measures and introduction of technologically advanced water saving equipment.

Finally based on the above results and analysis, the following external policy recommendations and internal actions are made for the case study production units to keep the total water footprint low as a measure of good environmental practice process and for achieving a sustainable future:

- Operational Blue WF constitutes the total water footprint, so the production units should aim for "Zero WF" or try to be "Water Neutral" which is compatible with sustainability goal. Industries should try to focus on reducing the water footprint at a given production level and by bringing down the evaporation and make up cycle loss. All of which involves huge financial requirement. Again if all the freshwater water that is extracted is returned to the nearby hydrological bodies in treated form after use in the production process, then blue WF can be reduced to zero.

  However the barrier to this possible alternative solution is that exiting national policy do not make Zero WF mandatory. A possible solution that can be suggested for breaking this barrier is that the norm prescribed by the Ministry of Environment, Forest and Climate Change can be changed and initiatives of providing subsidies to industries by the government can be taken. So industries can move towards zero WF if zero footprint is made mandatory by the Government along with providing suitable subsidy and industries can face the challenge to mobilize the financial requirement for achieving this much desired goal of zero WF.

Acknowledgements I thankfully acknowledge University Grants Commission (UGC), New Delhi, India for funding this work under Project Reference F. PHW-No. F.PHW-048/14-15 (ERO) Dated 03.02.2015.

Annexure-1

ANNEXURE -1

QUESTIONNAIRE

Corporate Water Use Accounting: Estimating Water Footprint of Select Indian Industry
sanctioned by UGC [MRP no. F.NO.PHW- 085/14-15 (ER0)].

Water Footprint Calculation

1. Annual output (in units) -

2. Sources from which water is used (in m3) by the Unit
    i) River water

3. How much of the ground water is not retuned hydrological system(in m3) from which it was withdrawn (i.e. evaporates or is incorporated in the products) ?

4. Do you support any plantation activities?

5. Do you supply water for carrying out any process of irrigation?

6. If yes, how much water is supplied(in m3)?

7. How many liters of water is used to produce steam?

8. Do you discharge effluent water? If yes, specify.

9. Do your unit have any drinking / wastewater treatment plant?(Y/ N) Y

10. If yes, how much electricity is consumed (in kWh) by that drinking / wastewater treatment plant?

11. Are you aware of the concept of water scarcity?

12. Did your company install water meters to measure water units?

    If Yes , Does your unit keep documentation of meter reading?

    If Yes , for past how many years such documentation is being maintained?

13. (a) Do your company adopt any water conservation measures?
       Can you give details of the water conservation measures undertaken by your unit ?

    For past how many years such conservation measures have been maintained?

    (b) If yes, please specify --

14. Consumption of water from two different sources for three different purposes (i.e. production , cooling process and domestic use)

	Freshwater consumption	Polluted water discharged
Process –m^3 / day Cooling –m^3 / day Domestic- m^3 / day (kichen, toilet, gardening etc.)		

15. How much cost has been incurred for water conservation / recycling /wastewater treatment etc.(in Rs).

16. How much cost reduction has been achieved as a result of adoption of Water conservation / recycling/ wastewater treatment mentioned measures?

17. How much investments have been made for the above mentioned activities?

18. Does your unit calculate water footprint ? (Y/ N)

19. If yes , from which year?

20. If yes, mention the WF (in m3) for the last few years.

21. Do your company (any unit of your company) support Rainwater Harvesting activities?

22. For how many years rainwater harvesting activity is being carried out ?

22. If yes, how much rain water harvesting has been made?

23. How much harvested rainwater is being used ?

24. Cost of undertaking Rainwater harvesting ?

References

Annual Survey of Industries-Factory Sector (Various Volumes). Ministry of Statistics and Programme Implementation, Central Statistical Office (CSO), Industrial Statistics Wing, Government of India.

Central Electricity Authority (CEA), CO_2 Baseline Database for the Indian Power Sector. User Guide, Version 10.0, Government of India, Ministry of Power, Central Electricity Authority. 1–32, December, 2014.

Chakraborty, D. (2012). *Performance evaluation of the Indian industries and environmental management practices*. Doctoral Thesis, Jadavpur University, Kolkata, India.

Chakraborty, D., & Roy, J. (2012). Accounting for corporate water use: Estimating water footprint of an Indian Paper Production Unit. *Indian Accounting Review, 16*(2), 34–42.

Chapagain, A. K., Hoekstra, A. Y., Savenije, H. H. G., & Gautam, R. (2006). The Water Footprint of cotton consumption: An assessment of the impact of worldwide consumption of cotton products on the water resources in the cotton producing countries. *Journal of Ecological Economics, 60*(1), 186–203.

Chapagain, A. K., & Hoekstra, A. Y. (2007). The Water Footprint of coffee and tea consumption in the Netherlands. *Journal of Ecological Economics., 64*(1), 109–118.

Ercin, E., & Hoekstra, A. Y. (2012). *Carbon and water footprints: Concepts, methodologies and policy responses*. (World Water Assessment Programme; No. 4). Paris, France: United Nations Educational, Scientific and Cultural Organization (UNESCO).

Evangelou, E., Tsadilas, C., Tserlikakis, N., Tsitouras, A., & Kyritsis, A. (2016). Water footprint of industrial tomato cultivations in the Pinios river basin: Soil properties interactions. *Water, 8* (11), 515. https://doi.org/10.3390/w8110515.

FAO. (2012).Coping with water scarcity. An action framework for agriculture and food security, FAO Water Reports: 38. Food and Agriculture Organization of The United Nations, Rome, 2012.

Gerbens-Leenes, P. W., Moll, H. C., & Schoot Uiterkamp, A. J. M. (2003). Design and development of a measuring method for environmental sustainability in food production systems. *Ecological Economics, 46*(2), 231–248.

Guzmán, C.D., Verzijl, A., Zwarteveen, M. (2017). Water footprints and 'Pozas': Conversations about practices and knowledges of water efficiency. *Water, 9*(1), 1–15. https://doi.org/10.3390/ w9010016.

Hoekstra, A. Y. (2017). Water footprint assessment: Evolvement of a new research field. *Water Resource Management, 31*, 3061–3081. https://doi.org/10.1007/s11269-017-1618-5.

Hoekstra, A. Y., & Chapagain, A. K. (2007a). The Water Footprints of Morocco and the Netherlands: Global water use as a result of domestic consumption of agricultural commodities. *Journal of Ecological Economics, 64*(1), 143–151.

Hoekstra, A. Y., & Chapagain, A. K. (2007b). Water Footprint of Nations: Water use by people as a function of their consumption pattern. *Journal of Water Resource Management, 21*, 35–48.

Hoekstra, A. Y., & Chapagain, A. K. (2008). *Globalization of water: Sharing the planet's freshwater resources* (pp. 1–232). London: Blackwell Publishing.

Hoekstra, A. Y. (2007). Human appropriation of natural capital: Comparing ecological footprint and water footprint analysis'. Value of Water Research Report Series No. 23 of UNESCO-IHE, Institute of Water Education, Delft, The Netherlands.

Hoekstra, A. Y., Chapagain, A. K., Aldaya, M. M., & Mekonnen, M. M. (2011). *The water footprint assessment manual: Setting the global standard*. London: Earthscan.

ISO 14046:2014 Environmental management—Water footprint—Principles, requirements and guidelines, Technical Committee: ISO/TC 207/SC 5 Life cycle assessment, 1–33, August, 2014.

Kampman, D. A. (2007). *Water footprint of India: A study on water use in relation to the consumption of agricultural goods in the Indian states*. Masters Thesis, University of Twente, Enscede, The Netherlands, 1–77.

Leenes, P. W., Hoekstra, A. Y. (2008). Business water footprint accounting: A tool to access how production of goods and services impacts on freshwater resources worldwide. Value for Water Research Report Series No. 27 UNESCO – IHE, Institute of Water Education, 1–46.

Ma, J., Hoekstra, A. Y., Wang, H., Chapagain, A. K., & Wang, D. (2006). Virtual versus real water transfers within China. *Journal of Philosophical Transaction of Royal Society London B, 361*(1469), 835–842.

Ministry of Environment , Forest and Climate Change for Thermal Power Plants, Part II, Sec 3 (ii) , Government of India, New Delhi, the 7th December, 2015.

van Oel, P. R., & Hoekstra, A. Y. (2012). Towards quantification of the water footprint of paper: A first estimate of its consumptive component. *Journal of Water Resource Management, 26,* 733–749.

Syamroy, M. (2011). *Accounting for air and water resources: A case study of West Bengal.* Doctoral Thesis, Jadavpur University, Kolkata, India.

Verma, S., Kampman, D. A., van der Zaag, P., & Hoekstra, A. Y. (2009). Going against the flow: A critical analysis of inter-state virtual water trade in the context of India's National River Linking Program. *Elsevier Journal of Physics and Chemistry of Earth, 34,* 261–269.

Walsh, B. P., Murray, M. N., & Sullivan, D. J. V. (2015). The water energy nexus, an ISO50001 water case study and the need for a water value system. *Water Resources and Industry, 10,* 15–28. https://doi.org/10.1016/j.wri.2015.02.001.

WBCSD. (2006). *Business in the world of water: WBCSD scenarios to 2025.* Conches-Geneva, Switzerland: World Business Council for Sustainable Development.

World Economic Forum (WEF). (2017). *The Global Risks Report 2017* (12th ed.). World Economic Forum: Geneva, Switzerland.

World Nuclear Association (WNA) Report. (2011). *Comparison of Lifecycle Greenhouse Gas Emissions of Various Electricity Generation Sources,* 1–12.

Zhang, Y., Huang, K., Yu, Y., & Yang, B. (2017). Mapping of water footprint research: A bibliometric analysis during 2006–2015. *Journal of Cleaner Production, 149,* 70–79.

Website

www.waterfootprint.org/tool/home/.
www.world.nuclear.org.

Water Footprint and Food Products

Ignacio Cazcarro, Rosa Duarte and Julio Sánchez-Chóliz

Abstract Water footprint (WF) analysis has been quite extensive in characterizing water contents in crops and animal products, which translated into processed products it covers an important spectrum of food products. The food industry then also interacts as supplier of goods and as demander of other inputs with many other sectors, e.g. notably agriculture and distribution sectors, but also with other less known sectors, such as the chemical sector, energy sector, etc. affecting its overall role in the economy. This agri-food activity becomes important (e.g. in several regions of Spain being the second higher in GDP after sectors related to the automobile industry), but also frequently due to its higher representativeness in employment and exports than other industrial sectors, etc. For example, in the EU the Agri-food sector accounts for more than 7% of the overall exports according to data of the most recent years. This is based mostly on crops such as cereals, but also transformed food industry products such as olive oil, wine, pasta, dairy products, meats, and processed products in general. In other regions we may see e.g. the important role of the US as importer, but also major exporter of grains, etc., the exporting role of India in milk (and ultimately then of water through these processes), or the high dependency of food (and hence WF from products abroad) of China. If all these processes are well captured with specific supply chains/process analysis, also studies on extended environmental input-output (IO) models complement such information by helping identifying the final sector of export or of distribution to the households, e.g. to what extent WFs occur from the consumption of wholesale or retail trade, or from activities such as hotels and restaurants. All these insights can be obtained from local to global models, and with more or less

I. Cazcarro (✉)
ARAID (Aragonese Agency for Research and Development), Department of Economic Analysis, Agrifood Institute of Aragon (IA2), University of Zaragoza, Saragossa, Spain
e-mail: icazcarr@unizar.es

I. Cazcarro
BC3-Basque Centre for Climate Change – Klima Aldaketa Ikergai, Bilbao, Spain

R. Duarte · J. Sánchez-Chóliz
Faculty of Business, Department of Economic Analysis, Agrifood Institute of Aragon (IA2), University of Zaragoza & Economics, Saragossa, Spain

© Springer Nature Singapore Pte Ltd. 2019   45
S. S. Muthu (ed.), *Environmental Water Footprints*,
Environmental Footprints and Eco-design of Products and Processes,
https://doi.org/10.1007/978-981-13-2454-3_3

detail in terms of products and sectors depending on the database used. We analyze these issues, providing as well general figures on the food products water footprint globally, but also going down into nations or regions. In that regard, we highlight the interest of some global food supply chains, both at global level and local level, its environmental relevance and the differences in estimations obtained from the different analytical methods. Our results show the high sectoral heterogeneity from the point of view of water uses and water footprint in the agro-food system and they also confirm the great differences that can arise with the level of disaggregation used, proving again that it is better more rather than less disaggregation in environmental information.

Keywords Water footprint · Crops to food supply chains · Food products Producer and consumer perspectives

1 Introduction

In this chapter, departing from the concepts on Water Footprint (WF),[1] we focus on the agri-food supply chains up to the final consumption (i.e., goods or services consumed by individuals or households for private consumption).

In other words, given the increasing importance of water footprint studies in the scientific literature and practice of water and agricultural management (Aldaya et al. 2010; Allan 1998; Chapagain et al. 2006; Dominguez-Faus et al. 2009; Hoekstra et al. 2009a, b, 2011; Hoekstra and Chapagain 2008; Postel 2000; Vörösmarty et al. 2015), we discuss the length and relevant boundaries of the agri-food system both at the local and global level. With this term, agri-food system, we refer to key supply chains linked to agricultural products before food consumption. This includes then the (agri-)food industries, but also notably sectors such as the trade and transport sectors, and as important final demand category, the hotels and restaurants sectors. Similarly, we will highlight to what extent sectors typically non-food users of primary products such as the paper industry, and wood & cork, relate to this agri-food system and how they become affected by the proper (or not) consideration of the specific agri-food supply chains.

Following this idea, we stress here the importance of understanding the differences in the methods used for the computation of direct water use and embodied (virtual) water. Among the databases and methods so far used, they tend to differ in the interest of studying quantitative absolute or relative volumes or pressures, and also importantly on the selection of boundaries of the supply chains (whether shorter but more precise, or longer, but in a cruder form). We may highlight the

---

[1]The water footprint is an indicator of freshwater use that looks at both direct and indirect use of water by a consumer or producer. The water footprint of an individual, community or business is defined as the total volume of freshwater that is used to produce the goods and services consumed by the individual or community or produced by the business.

bottom-up methods based on specific very detailed (in terms of crops and their transformation) supply chains, life cycle assessment (LCA) methods, or the top-down ones such as Input-output (IO) databases and models. In this last line of research, important advances in the last decade have occurred regarding the development and use of Multi-regional input-output (MRIO) for trade analyses and for the study of environmental challenges. A few of them have also focused on detailing the environmental key sectors, so that the environmental analyses (on water, carbon, materials, etc.) are richer.

In this regard, we focus this chapter on the water footprints related to the final demand of food products, and how these estimates may change depending on the structure of the databases in the agri-food system. More precisely, and following some referential works which advocate for trying to keep more rather than less disaggregation in environmental information (Lenzen 2011), we examine the relevance of the aggregation bias when the water analyses are performed with crucial activities such as agriculture aggregated in a single or few categories.

The remainder of the chapter is organized as follows. The second section deals with a literature review focused on the methods and estimates of water footprint of food products. The third section focuses more on the differences and complementarities of the approaches, which we consider particularly suited for the estimates of agri-food products, and insights on their integration. In the fourth section we present three case studies: the results obtained for the global economy with the EXIOPOL database and for two models at a lower geographical scale, one from the Spanish economy and other from Huesca, a Spanish region. In both cases the high sectoral heterogeneity from the point of view of water uses and water footprint in the agro-food system is revealed. We also confirm the great differences that can arise with the level of disaggregation used, proving again that it is better more rather than less disaggregation in environmental information. Building upon the identified difficulties and promising prospects for analysis, the final discussion drafts open and future lines of research on water footprint of food products.

2 General Literature Review on Water Footprints Associated to Food Products

The defined water footprint indicator grew as other ones on pressures, given that human societies use huge amounts of water, with increasing competition for scarce resources, also impacting on the present and future state of the environment. Furthermore the concept grew quickly in importance with the publication of numerous papers, the celebration of conferences and sessions on the topic, and the establishment of an international Water Footprint Network (WFN). The concept originated linked to that of "virtual water"[2] coined by (Allan 1993, 1994, 1996, 1998), also called "embedded/embodied water" or "hidden water".

We may refer to the Water footprint manual (in their different versions (Hoekstra et al. 2009a, b, 2011) as a main reference of the water footprint concepts and studies, clearly focused in the agri-food supply chains, gathering the most popular and cited concepts, methodologies, etc. Also we may cite some of the comprehensive reviews of water footprint studies (Chenoweth et al. 2014) and (Zhang et al. 2017). We may cite the reviews of (Chenoweth et al. 2014; Daniels et al. 2011; Duarte and Yang 2011) as comprehensive reviews of top-down approaches, while the question of comparison of these approaches with those bottom-up has been studied conceptually in Chenoweth et al. (2014), Yang et al. (2013) and empirically e.g. in Feng et al. (2011).

We focus now more specifically on the importance for the study of water footprints of food products. In any case, most of this type of studies (except for the threads of literature on forest and paper products, textiles, and on bioenergy) already have the focus on the agri-food supply chains, leading mostly to food products, being aware of the fact that these chains are the ones with more embodied (virtual) water. Examples of these are (Hoekstra and Hung 2002) and many subsequent studies such as (Chapagain and Hoekstra 2003, 2004; Hoekstra and Chapagain 2007; Mekonnen and Hoekstra 2010, 2011a). Hoekstra (2013), Mekonnen and Hoekstra (2014) proposed benchmarks for fair water footprint shares, and formulated water footprint reduction targets, being this last objective also found in Lutter et al. (2014), Lutter and Giljum (2015). Together with this growing relevance and enthusiasm (Verma et al. 2009) discussed the importance of non-water factors in determining trade across and between countries, and (Gawel and Bernsen 2011, 2013; Perry 2014; Wichelns 2010, 2011a, b, 2015; Witmer and Cleij 2012) performed critical evaluations of the water footprint concept. We omit here an exhaustive long review of all those type of studies summarized above, especially related to the WFN, since most literature reviews on virtual water and water footprint already address them.

What we may indicate is that apart from academia, the water footprint concept has received increasing press coverage, and a growing number of countries, businesses (Coca-Cola and Nature Conservancy 2010; Cooper et al. 2011; Pepsi-Co 2011) and organisations (WWF 2012) moved towards quantifying aspects of their operations related to water, using the water footprint (WFN 2018). As somehow studies introducing some differences in the way of computing water footprints, we may also highlight (SABMiller_WWF-UK 2009) using the concept of the net green water footprint, and (Herath et al. 2011) proposing a net water balance water footprint, applied to hydroelectricity and kiwi fruit production.

The International Standardization Organization (ISO 2014) considered developing a new international standard for water footprinting in order to complement its

---

[2]Virtual water referred to the volume of water required to grow, produce and package agricultural commodities and consumer goods.

existing Life Cycle Assessment (LCA) standard (Humbert 2009; ISO 2017). Precisely this literature of LCA, which also can be seen as bottom-up approach, but not so distant from top-down approaches when the hybrid methods are used, has also been relevant for the development of the concept. The LCA has approached to the water footprint concepts (Bayart et al. 2010; Berger and Finkbeiner 2010, 2012; Boulay et al. 2018; Hoekstra et al. 2009a, b; Kounina et al. 2013; Milà i Canals et al. 2009; Pfister et al. 2017; Ridoutt and Pfister 2010), although its attention has often been paid more to the desirability or not of synthetic indicators of the water footprint, to reflect the impacts of production systems and consumption patterns, etc. In this regard, also studies such as Vanham and Bidoglio (2013) suggested an extended the analytical framework that includes sustainability assessments by combining social and economic factors with water footprints.

In the case of top-down approaches, such as the extended environmental input-output (IO) models, the share of studies which have put specific emphasis in computing specific and more insightful agri-food chains, or/and that have compared conceptually or empirically their results with other methods, is much smaller compared to the total number of environmental IO studies (being these other with interests such as net trade, water footprint per capita, water use changes, etc. Still, we may cite quite a number of them with agri-food system focus: (Bogra et al. 2016; Cazcarro et al. 2010, 2012, 2014; Lenzen 2009; Lenzen et al. 2012a, b, 2013a; Lenzen and Foran 2001; Lutter et al. 2013; Stadler et al. 2016; Wood et al. 2015).

As we will see in the following section, going deeper into the methods, the main differences among them typically have to do with the system boundaries selection, which in essence archetypally involve truncation errors on the one hand (bottom-up methods), and aggregation biases on the other (top-down).

3 Methods for Agri-Food Analysis, and Proposals of Integration

3.1 The General Issue of System Boundaries

Our point of departure to explain the reasons for the different methods proposed for accounting virtual water and water footprints has to do with the question of truncation errors and the system boundaries selection (Ekvall and Weidema 2004; Lenzen and Treloar 2002, 2003; Suh 2004; Treloar 1997), which has had a long lasting literature, but probably finds a landmark or referential works in Suh et al. (2004) and Suh et al. (2009) (see particularly Ferrão and Nhambiu 2009; Lifset 2009). One of the main insights from these works is that the standards by the International Organization for Standardization (ISO) which are often followed in LCA studies, impose practical difficulties for drawing system boundaries. Decisions on inclusion or exclusion of processes in an analysis (the cut-off criteria) are typically not made on a scientific basis. In particular, deciding which processes could be excluded from the inventory can be rather difficult to meet because many excluded processes have often never been assessed by the practitioner, and therefore, their negligibility cannot be guaranteed.

Then the subjective determination of the system boundary may lead to invalid results. In this regard, LCA studies utilizing economic input-output analysis have shown that, in practice, excluded processes can contribute as much to the product system under study as included processes. For example an interesting recent application is found in (Hauke et al. 2017). It contrasts the LCA truncation errors in different frameworks using the same underlying IO data set and varying cut-off criteria. The work shows that modelling choices in an LCA process can significantly influence estimates obtained and that differences in IO and process inventory databases, such as missing service sector activities (e.g. of trade, information, finance and insurance, businesses activities, etc.), can significantly affect estimates of LCA truncation errors.

A counterargument to these last conclusions would be the lack of sectoral/ product disaggregation or detail in the intensities accounted for. For example, with an IO which considers all together a single sector of "agriculture and forestry", we may find higher than real embodied (virtual) water intensities, and hence embodied water in the final purchases, of products such as paper or wood and furniture. For this reason, we consider that in order to perform fairer comparisons in that regard studies which are used as benchmark should have a higher disaggregation. This relates to an important literature in IO studies on the "aggregation bias" (Bouwmeester and Oosterhaven 2013; Crown 1990; Jalili 2005; Kymn 1990; Lenzen 2011; Morimoto 1970, 1971; Olsen 2000).[3] Although we will devote much less attention to the "spatial aggregation bias" (Lahr and Stevens 2002; Su and Ang 2010), since we consider that it has been less determinant in the choices of methods, it is also worth considering. We will address these issues in practice later on.

3.2   The System Boundaries for Water Footprints

In the case of water footprints, the intuitive main sectors/products and water flows can logically be thought to be the agri-food supply chain, and a few other related ones such as textiles, paper or bioenergy, which are the ones that the studies of the WFN have especially accounted for. The intuition or insight obviously is founded and grounded in data, notably of water uses and of intersectoral and international trade of products. For many countries globally the main water user/consumer is agriculture (which in many IO tables is grouped all together in a single account,

---

[3]In relation to other environmental pressures or impacts, we find comprehensive reviews of the literature in Lifset (2009) for CO_2 emissions, and in de Koning et al. (2015), Pittel et al. (2012) for raw materials.

including also forestry), and in particular the crops production (about 60–70% of
the water, depending on the metrics used). For this reason, it is probably fair to say
that those type of WFN studies are great in capturing the lion's share of water uses
and flows in the world, with great agri-food detail and heterogenous water intensities. We may roughly represent this with the size of this sector/products in Fig. 1.
If we separate the crop production (called "Agriculture and forestry" in Fig. 1) from
the animal categories (called "Livestock"), as also several original accounting IO
frameworks do, we would find a large flow of materials, and hence of virtual water,
from "Crop production" to "Livestock".

In the case of "livestock" itself, the size of the oval representing the water uses
directly (on site, not via "virtual" water from crops and other inputs) may be not
insignificant, and particularly more if green water from pastureland is accounted for
and attributed to this "livestock" category (contrary to other fodder crops, pastureland typically would not be considered in an IO framework, except if, as we
propose, it is added for the environmentally extended framework). The food
industry clearly interrelates with these sectors/products, given that it typically is an
important buyer of crop and animal products, which often are also later distributed
via trade, transport and distribution sectors. But also, the food industry sector it is
particularly important as an intermediate node between the crops production and
livestock consumption, by having an important sector of industrial feed processing.

If we had to highlight one single account which the WFN approach may have
been less able to fully capture, we would highlight the production and distribution
of electricity and gas (which we may also call in a general form "Utilities").

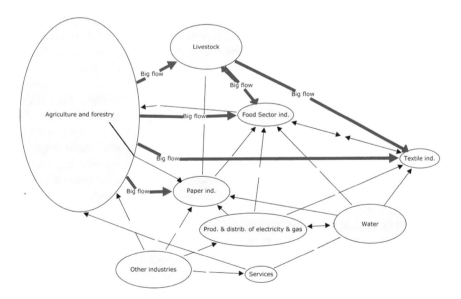

Fig. 1 Schematic representation of the main direct flows among economic sectors/products.
Source Own elaboration

The pressures of this sector however have always been under debate, given the different metrics which are considered (water withdrawals, blue, green or grey water consumption, etc.), given that it used to be considered that its "actual" physical consumption is not so much (the non-consumptive part of water withdrawals, i.e., the return flow, is not part of the water footprint). In other words, it has been debated whether hydroelectric generation is merely an in-stream water user or it also consumes water (Aguilar et al. 2011; Cooley et al. 2011). To provide arguments for the importance of its consideration (see Mekonnen and Hoekstra 2011b) arguing that hydroelectric generation is in most cases a significant water consumer), one may highlight the importance of water evaporation from lakes and other water surfaces, whose purposes are often the electricity production, or in combination to water delivery for agriculture and other demands.

One may also highlight the fact that often in thermal energy uses the heating of water is important for several ecological functions, and so that measures of water footprints should account for this, in the same fashion that grey water footprint does not measure water volumes "used" but "needed" for assimilation. Similarly, we may also stress other aspects about the role of the sector of water distribution. Even though measures of water footprint focus on consumption, typically this sector often also distributes water to agriculture (and obviously to other sectors and to households), incurring in other losses (inefficiencies, etc.) which are relevant for water management. We will further look at the results from these sectors in the following section with the empirical results.

It should be highlighted that in schematic framework above we have not shown other important sectors in the supply chains, as notably the trade and transport sectors, and as important final demand category, the food obtained from the hotels and restaurants sectors. Each of those steps and sectors also have some (relatively small) water uses, which add to the global accounting (in Hoekstra et al. 2009a, b, 2011) is accounted as "water for processing"). To the schematic framework above, we now also need to add the fact that the interaction among all these sectors and also among many countries, via trade, and so the ability to capture the embodied (virtual) water flows among sectors/products. In this regard the water uses intensities need to be heterogeneous enough to account for those regional/country differences. In general trade statistics such as UN Comtrade (2018), widely used in most studies associated to the WFN for the computation of "virtual water imports" and "virtual water exports", are highly detailed in terms of products but also of countries (close to 200). MRIO type of databases tend to have country detail, particularly for the richest countries, leading to around 40 regions in WIOD (Dietzenbacher et al. 2013), 48 in EXIOBASE (Tukker et al. 2009, 2013; Wood et al. 2015), around 65 in OECD (2016), 140 in GTAP (Narayanan et al. 2012, 2015) and close to 190 in EORA (Lenzen et al. 2012c, 2013b). But crucially, apart from using the same type of original information from UN Comtrade (Eurostat 2016; Stehrer and Rueda-Cantuche 2016), these second type of studies show important advantages, e.g. typically trying to make consistent bilateral trade flows and properly framing the distinction between intermediate goods and services and final demand ones. In this regard, "virtual water imports" and "virtual water exports" are not obtained from a kind of bilateral trade statistics which treat all products equally (as final) and hence with the water (footprint of production) intensities from the source country, but accounts all the water involved through the production processes (e.g. of a cereals produced in a country, which are transformed for feed for animals in the food industry of a second country, which are sold be provided to animals of a third country, which furthermore can be scarified to obtain leather in a fourth country, to be consumed in a fifth country, etc.). Related to this, the final demand of products is differentiated with respect to the region of origin, and so the food bought in a country would have different embodied water intensities (multipliers in the IO framework) depending on whether they have been finally produced in the country itself or elsewhere.

Finally, at all moments we have discussed boundaries in terms of sectoral and regional detail. One could refer also at the boundary or extent to which the analyst wants to reach with the indicator, regarding measuring, or highlighting/providing evidence on pressures, whether in absolute terms (water volumes) or in relative terms, etc. In that regard (Pfister et al. 2017) aimed to explain the role and goal of LCA and ISO-compatible water footprinting and claimed to resolve the six issues raised by Hoekstra (2016). By clarifying those concerns, they identified the overlapping goals in the WFN and LCA water footprint assessments and discrepancies between them. The main conclusion was that the LCA-based approach aims to account for environmental impacts, while the WFN aims to account for water productivity of global fresh water as a limited resource.

Also in Cazcarro and Arto (2018) it is indicated that hybrid approaches could be used, aiming to profit from the best of LCA and IO worlds, but in the following section of results we will focus on the results from MRIO on water footprinting, and to what extent the level of detail is important. The following subsection deals with the different data discrepancies among MRIO databases and the reasons for choosing one in the analysis of this chapter.

3.3   Data Discrepancies and Choice of the Global Database

As indicated in Cazcarro and Arto (2018), the water accounts of EXIOBASE (Tukker et al. 2009, 2013; Wood et al. 2015) reveal quite notable differences with respect to other works. Making use of WIOD (Dietzenbacher et al. 2013) and the associated water accounts (Arto et al. 2012, 2016), from 1995 to 2009 the global water footprints ranged annually from 8,700 to 12,000 km^3 (representing about 70% green, 17% blue and 13% grey). In particular for the year 2007, close to 8,000 km^3 was green water and 1,900 km^3 of blue water footprint of production. If we look at EXIOBASE version 2.2. (Tukker et al. 2009), we find that about 4,480 km^3 is green (4,423 km^3 of crops, and 0,039 km^3 of animals' categories), while about 1,600 km^3 is blue water footprint of production. Although both databases depart from the main water footprint studies, such as (Mekonnen and Hoekstra 2010, 2011a, b), we find discrepancies between them, since in those last studies the water footprint of crops represents about 5,000 km^3, and the water footprint of animals and animal products represent more than 2,100 km^3. Certainly, there should recognize that in EXIOBASE only the direct water consumption by animals is taken into account (while water footprint from feed comes from the purchases of those animal categories to the crops categories, from the processed feed of the food industry, etc.), but also alternative possibilities with respect to animals' green water consumption computation exist. Despite this fact that we see as a drawback of using EXIOBASE, in any case in terms of structure of the database, i.e. having a large sectoral and environmental detail with a common classification of sectors for all regions, while keeping an important regional detail (not as much as EORA, but larger than WIOD, and still with common sectoral classification, and with publicly available water accounts, which is not the case e.g. of GTAP), we consider that it is the most appropriate for our purposes. Regarding these databases we should also indicate that EORA provides the same global volumes for all the period 1990–2013, estimating a total of 7,601 km^3, being 341 km^3 grey water and the rest estimates of blue and green water footprint of production from crops and all the other sectors. A database such as EORA also allows for multiple analysis of aggregation bias, given that it aims to keep the maximum disaggregation of all tables it gathers or estimates, but we considered that it did not served us for the purpose of having an initial general comparison which might be done in the same form for many countries. With respect to GTAP, although water accounts have been linked to the database (see e.g. Cazcarro et al. 2016; Haqiqi et al. 2016; Taheripour et al. 2013; Tol 2011; Tol et al. 2011), no official, publicly available satellite water accounts exist.

3.4 Data Description

As we have said, we want to approach our analysis at two levels, global and local. For the first ones, we use here then EXIOBASE version 2.2 for 2007 that provides data for an extended environmentally multi-regional input-output (EE-MRIO) model for 163 industries, 200 products, and 48 countries and regions. The detail in agriculture is 8 categories of crops, 7 of animals (plus 2 on manure), it also has a category of forestry, and another one of fisheries. In the case of the food industry, there is a distinction of 10 categories, plus one of beverages and one of tobacco industry. In trade no distinction between food products is performed, and also one single category of Hotel and restaurant services is registered.

Regarding the regional detail, it comprises 48 countries or world regions, more precisely 27 EU countries, 16 non-EU countries and five regions. Single countries considered (43) cover 90% of global gross domestic product (GDP). For this study 'industry by industry' tables were chosen due to the classification by industries of the water data. However, due to the interest in products in this book, we also establish a simplified correspondence of industries to products for the water accounts, and compute the embodied (virtual) water of products also with the 'product by product' table, of 200 products times 48 regions.

Regarding the water accounts, in the course of the CREEA project (EXIOBASE 2), real data on water withdrawal and consumption was found to be scarce—especially on the international level- and modelled data (Lutter et al. 2014) was used (Stadler et al. 2016). As explained in the 4th Supplementary Material of (Stadler et al. 2018) (i.e. of the still not publicly available EXIOBASE 3), for the compilation of the water use/consumption extensions for the EXIOBASE set up and compiled in the EXIOPOL and CREEA projects, the main data sources used were the ETH dataset (Pfister et al. 2011; Pfister and Bayer 2014) and the Water Footprint dataset (Mekonnen and Hoekstra 2011a) for agricultural water consumption and the WaterGAP model (Flörke et al. 2013) for industrial water use/consumption. "These databases are currently among the most comprehensive global databases with the agricultural water consumption datasets encompassing a vast number of agricultural categories and the WaterGAP data set covering a large number of livestock categories as well as manufacturing sectors—the latter being an area where special requirements of an MRIO system meet the general poor data coverage situation. In the update from DESIRE (it is also another European Union projects in line with EXIOPOL and CREEA, for the development of EXIOBASE), the two basic data sources used were the Water Footprint dataset (Mekonnen and Hoekstra 2011b) for agricultural water consumption based on FAO data and the WaterGAP model (Flörke et al. 2013) for industrial water use and water consumption".

For the second type of analyses, we use Social Accounting Matrices (SAMs, an extension of the IO tables) for Spanish province of Huesca and for the whole Spain (Cazcarro et al. 2010, 2014). The first one is a highly agricultural (crops and livestock) disaggregated table for the Spanish province of Huesca (Cazcarro et al. 2010). The second one is a database for the whole Spain, relatively less disaggregated in crops but more on the downstream supply chains of agriculture (Cazcarro et al. 2014). In both cases we will show two types of water accounts, water withdrawals on the one hand, and physical consumption (blue and green) on the other hand. In order to avoid mixing here all the details on sectoral disaggregation, etc., we provide the specifics together with the analysis of results and their discussion in Sect. 4.

4 Results and Challenges in Estimating Water Footprint of Agri-Food Products

4.1 Exploration of the Aggregation Bias and Related Issues with EXIOBASE

As hinted above, one of the main challenges we find for IO and MRIO databases is the aggregation bias when allocating the use of water of one country. The implicit assumption of equal water intensities (per unit of output) across all grouped products (e.g., agriculture, fishing and forestry in the OECD tables, agriculture in WIOD, or even in EORA for some countries) may lead to miscalculations. We then find further effects downstream of the supply chains, when the water content of one unit of output of the agriculture, fishing and forestry sector is implicitly assumed to be the same irrespective of the industry buying their agricultural products (e.g. as if the food industry was merged with the wood or paper industry). In order to explore these aspects, we evaluate the water footprint of consumption (and hence of products embodying water for final consumption) with EXIOBASE placing the focus on the agri-food system products. In Fig. 2 we summarize the ranking of activities of the (20 out of 200) embodying the largest water volumes (hm^3) for final consumption for different water types (green, blue, blue "scarce", withdrawal, withdrawal "scarce") for all countries. We may observe that largest ones are the WF of products from the food industries, followed by other primary crops. The activities that seem less related to food involving large water embodied (virtual) contents are construction, and Public Administrations (PP.AA.) and related activities such as health and social work services.

This is particularly noteworthy when looking at water accounts or types such as blue WF, or such as withdrawals. In this regard, in Table 1 we may see in the first column of values the ranking of the larger blue water footprints of consumption (by final demand category and country), production, and difference among the two (by activity). Following with the hints in Fig. 2, we may observe the different role of the activities in terms of WF of production (obviously very high in primary crops such as paddy rice, wheat, oil seeds, etc.) and WF of consumption (as described, very high in products from the food industry, such as "Food products nec"). In the case of activities from the food industry, but also other ones such as construction and PP.AA. often reveal very low WF of production, especially in relation to the WF of consumption. In this case we may also see that the largest WF of consumption are found in countries such as India (IN), United States of America (US), China (CN), Japan (JP) and 'Rest of the World Asia and Pacific' (WA).

In summary, the data show us that, at a global level, sectoral or geographical aggregation levels can be very relevant and change a lot the WF estimations, especially when we look at the agri-food sector with a broad scope. This forces us to look for the greatest disaggregation possible in our analyses and to carefully select the databases that we use.

4.2   Further Exploration of the Aggregation Bias with Agri-Food Detailed (Social Accounting) Matrices for Spain

The above analysis sets the grounds for the uncertainties, aggregation biases, etc. for an MRIO case with an important focus on the relevant sectors for environmental

Table 1 Ranking (40) largest blue water footprints (hm^3) of consumption (by final demand category), production, and difference (by activity)

CTRY	Activities	WF cons. (C)	WF prod. (P)	WF P-C
IN	Paddy rice	54,992	71,115	16,123
US	Food products nec	53,050	49	−53,000
IN	Wheat	49,805	52,808	3,003
IN	Products of vegetable oils and fats	38,974	131	−38,843
IN	Oil seeds	37,501	112,129	74,628
CN	**Construction work**	35,583	0	−35,583
CN	Food products nec	34,095	361	−33,734
IN	Vegetables, fruit, nuts	23,177	40,487	17,309
CN	Vegetables, fruit, nuts	22,839	47,531	24,691
WA	Hotel and restaurant services	21,341	20	−21,321
WA	Vegetables, fruit, nuts	20,288	41,643	21,355
IN	Cereal grains nec	20,266	18,852	−1,413
US	Vegetables, fruit, nuts	19,687	15,827	−3,860
TR	Food products nec	18,628	98	−18,530
IN	Crops nec	16,547	27,391	10,844
US	**Public administration and defence services compulsory social security services**	14,889	1,316	−13,573
IN	Food products nec	14,812	130	−14,682
WA	Wheat	14,782	68,409	53,627
CN	Oil seeds	14,171	30,706	16,536
CN	Fish products	13,378	370	−13,008
WA	**Construction work**	13,376	31	−13,344
JP	Food products nec	12,998	28	−12,970
CN	Processed rice	12,527	25	−12,502
CN	Cereal grains nec	11,945	32,256	20,311
MX	Food products nec	11,843	5	−11,837
WA	Food products nec	10,366	239	−10,127
IN	Hotel and restaurant services	10,226	0	−10,226
CN	**Health and social work services**	10,113	0	−10,113
WF	Food products nec	9,816	77	−9,740
WA	Crops nec	9,729	29,277	19,547
WA	**Public administration and defence services compulsory social security services**	9,496	146	−9,350
WM	Food products nec	9,333	7	−9,326
CN	**Public administration and defence services compulsory social security services**	8,536	0	−8,536
US	Beverages	8,447	26	−8,422
CN	Hotel and restaurant services	8,431	0	−8,431

(continued)

Table 1 (continued)

CTRY	Activities	WF cons. (C)	WF prod. (P)	WF P-C
CN	**Machinery and equipment n.e.c.**	7,916	2,330	−5,587
WA	Paddy rice	7,745	39,940	32,195
IN	**Other land transportation services**	7,696	0	−7,696
US	**Construction work**	7,689	0	−7,689
WA	**Health and social work services**	7,638	18	−7,621

Note nec stands for "not elsewhere classified"

In bold red we highlight activities that we would not consider as being within the "agri-food" system

IN India, *US* United States of America, *CN* China, *WA* 'Rest of the World (RoW) Asia and Pacific', *TR* Turkey, *JP* Japan, *MX* Mexico, *WF* RoW Africa, *WM* RoW Middle East

Source Own elaboration

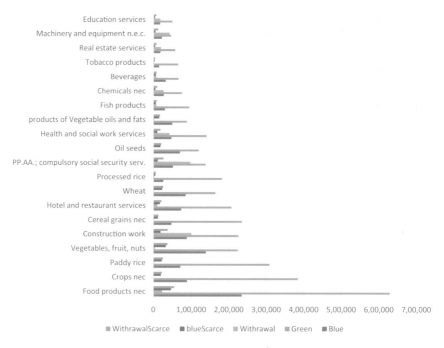

Fig. 2 Ranking of the 20 largest water footprints (hm^3) of consumption (by final demand category): green, blue, blue "scarce", withdrawal, withdrawal "scarce" for all countries., Note: The ranking is elaborated based on the blue + green water footprint of consumption. *Source* Own elaboration

analyses. Probably generalized further disaggregation in the agri-food system at such a global scale is not feasible nowadays. We may want to also look at the specific effects of disaggregation for single country analyses. Omitting the detail of all the assumptions, many possible critiques of single country analyses (see e.g. Jesper et al. 2008; Lenzen et al. 2004; Miller and Blair 2009), we explore the issues of aggregation for water footprinting of consumption with two additional databases (SAMs).

We may simplify the structure of the tables for Huesca and for Spain as below in Fig. 3, highlighting the crops and animal categories. In the case of the SAM for Huesca this means 32 and 9 livestock categories, while in the case of the SAM for Spain this means 6 crops categories and 4 livestock or animal categories. In order to understand where the main flows of the agri-food products and services occur, we may highlight in red the main origins and destinations of crops. In yellow, we also highlight relevant relations of animal categories (obviously in the interaction with crops the cells could have been represented as red as well). Important industrial processes that loop back into livestock are the yellow cells shown by the purchase of crops by the food industry, in particular the industry of feed (which we have as separate sector), which is then sold to the animal categories. In purple other relevant interactions, notably with agri-food sectors (such as the purchases by agriculture of energy, chemicals, etc.), being in blue less important ones for our purposes (agri-food and hence water flows), and in turquoise blue those among all other

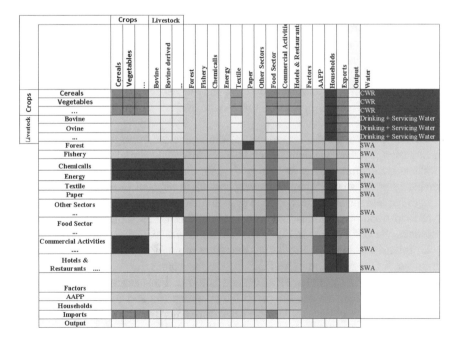

Fig. 3 Simplified structure of the Social Accounting Matrix for Spain, colouring flows according to the found importance in the agri-food system. *Source* Own elaboration

economic sectors. The green square represents the extension of the Social Accounting Matrix with monetary flows among institutional accounts. In the column of water we indicate that for crops the "Crop Water Requirements" (CWR) were considered, while for animals only the "Drinking and servicing water", since no accounting of green water (from pastureland) was taken. Water uses for other sectors are obtained from the Satellite Water Accounts (SWA) of the National Statistics Institute.

For simplicity, we examine the results of embodied (virtual) domestic (i.e. Huesca/Spanish respectively) water in the final demand (domestic and foreign, i.e. exports) of food related goods and services obtained from the disaggregated table and the table aggregated in one single agricultural sector.

In the first case of the SAM for Huesca, with a very high agricultural disaggregation, especially of crops, we compare now the results of this "disaggregated" framework and the "aggregated" version into 1 single agriculture category. We may see in Table 2 that in terms of absolute values, as one could expect, the largest differences are found for the aggregated sectors themselves. In particular, for water withdrawals 159 hm^3 (17% of change with respect to the disaggregated figure) of water more is found in final demand of agriculture with the aggregated table.

On the contrary, 110 hm^3 (13%) of water less is found in final demand of food industry with the aggregated table, and also 32.7 hm^3 (20%) of water less for hotels and restaurants final demand. Also we may observe notable differences of 9 hm^3 (46%) less in Paper, Publishing & Printing, and 6.5 hm^3 (22%) less in Public Services. On the contrary 3 hm^3 (a remarkable 126%!) more in Wood & cork is found in the computation with the "aggregated" vs. the "disaggregated" table. Similar direction and sizes of changes are found for the analysis of Physical consumption (blue and green). We may highlight that the sectors of "Construction & Engineering" and "Commercial services" show an opposite signs s with respect to the analysis of water withdrawals, and also the differences for "Energetic products" represent a much more important share (with respect to the disaggregated table value).

Having understood the importance of the agricultural disaggregation, but also having seen that the final demand, and hence embodied (virtual) water of agri-food products concentrates on a few categories, the final analysis is performed with a database for the whole Spain, relatively less disaggregated in agriculture (still 10 categories) and more on the agri-food system sectors, namely 16 in the food sector, 8 of food trade, 4 of hotels and restaurants and other 29 industrial and services accounts. We first provide in Fig. 4 the ranking of the 35 higher embodied (virtual) water intensities, highlighting before the name of the account in the general classification of agriculture (AGR.), Food, beverages and tobacco industry (FOOD_BEV_TOB.), trade (TRADE) and hotels and restaurants (HOT&RES.).

We may see that we need to reach the positions (33 and 35 ranking from the bottom of highest intensities to top, 3 and 1 ranking from the top) of embodied water withdrawals (which is the variable driving the ranking) to find relevant intensities out of these categories ("Manufacture of leather and footwear" and

Table 2 Difference in volumes and in percentage changes of embodied water with the aggregated table versus the disaggregated one. SAM for Huesca

	Water withdrawals		Physical consumption (blue and green)	
	Difference in Dm^3 (Agg-Disagg)	% wrt Disagg	Difference in Dm^3 (Agg-Disagg)	% wrt Disagg
Agriculture	159,062	17	141,785	32
Energetic products	−419	−3	−326	−22
Water	−5	−1	−5	−1
Metallic—Mineral products	−3	−3	−3	−7
Non-metallic—Mineral products	−42	−4	−41	−11
Chemical products	−198	−1	−199	−1
Machinery and equipment	−294	−4	−304	−8
Transport equipment	−43	−3	−44	−6
Food products, beverages and tobacco	−109,855	−13	−97,685	−20
Textile products, leather and footwear	−848	−15	−753	−23
Paper, Publishing & Printing	−9,212	−46	−7,618	−58
Wood & cork	3,069	126	1,421	91
Rubber and other plastic products	165	2	15	0
Construction & Engineering	778	5	−603	−7
Other manufacture, recycling and recovery	0	−3	0	−5
Commercial Services	−163	−2	161	5
Hotels & Restaurants	−32,710	−20	−27,969	−28
Transport & Communications	−87	−6	−86	−17%
Credit & Insurance	−95	−14	−86	−25%
Real estate and renting activities	−248	−9	−336	−22%
Private Education	−213	−17	−184	−27%
Private Health activities	−436	−14	−372	−21%
Other sales-oriented services	−375	−12	−282	−17%
Domestic service		−		−
State Education	−241	−17	−210	−27%
Public Health	−1,217	−12	−1,062	−19%
Public Services	−6,369	−22	−5,213	−27%
Total	0	0	0	0

Source Own elaboration

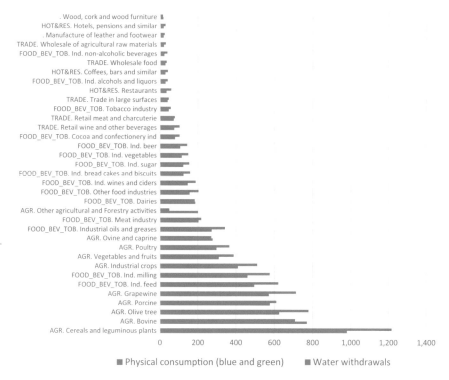

Fig. 4 Embodied (virtual) domestic water intensities (m³/1000 € of final demand) by ranking (top 35). *Source* Own elaboration

"Wood, cork and wood furniture"). It becomes clear then the quite heterogeneous intensities, and the biases that might occur when considered in aggregated forms.

This might be somehow masked when one just looks at the absolute values, as we do in Fig. 5, where the embodied (virtual) water volumes (hm³ by final demand categories) by ranking are shown. The magnitude of the final demands has a major importance, and hence here we find high volumes in other sectors already in the ranking (from the bottom with the highest volumes) 13 and 14 ("Construction and engineering" and "Other services for sale"). The ranking is also interesting in reflecting how most water volumes are embodied in vegetables and fruits, which are of high consumption in Spain, but also quite importantly exported to many places in the world.

We compare now the results of this "disaggregated" framework and the "aggregated" version into 1 single agriculture category, a category of the food sector, another of trade, and another of hotels and restaurants. We show this in Table 3 (like Table 2) by showing the absolute and percentage differences by product.

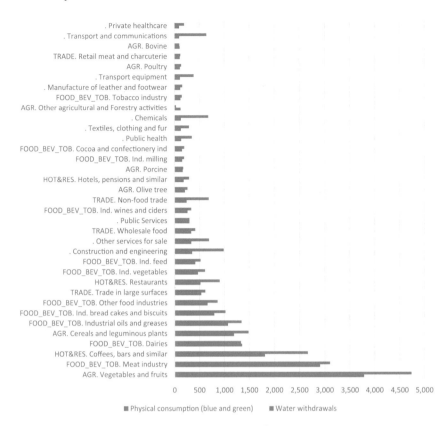

Fig. 5 Embodied (virtual) domestic water volumes (hm^3 by final demand category) by ranking (top 35). *Source* Own elaboration

We may see that in terms of absolute values, as one could expect, the largest differences are found for the aggregated sectors themselves. In particular, for water withdrawals 1,631 hm^3 (28% of change with respect to the disaggregated figure) of water more is found in final demand of agriculture with the aggregated table (probably the larger size of the intrasectoral purchases makes larger volumes of water to be attributed to the products purchased by final demands). On the contrary, 2,101 hm^3 (25%) of water less is found in final demand of food industry with the aggregated table, as well as 229 hm^3 (18%) less in the case of trade. It should be noted that the direction of differences between the aggregated and disaggregated versions are the same ones that those shown for the SAM of Huesca. But contrary to that table where the trade category almost did not reflect any difference, here the capability of distinguishing the trade of agricultural products makes the differences of the computation of water volumes from the aggregated and disaggregated tables reveals an important difference in this aggregated category of trade.

Table 3 Difference in volumes and in percentage changes of embodied water with the aggregated table vs. the disaggregated one. SAM for Spain

	Water withdrawals		Physical consumption (blue and green)	
	Difference in hm^3 (Agg-Disagg)	% wrt Disagg	Difference in hm^3 (Agg-Disagg)	% wrt Disagg
Agriculture	1,630,698	28	409,870	4
Extraction of energy products	4	1	10	0
Coking, refining and nuclear fuels	1,470	4	2,775	0
Production and distribution of electricity and gas	1,399	6	3,188	0
Water	933	5	1,865	0
Minerals and metals	3	1	7	0
Minerals and non-metallic mineral products	3,237	22	5,850	8
Chemicals	−2,586	−2	15,668	2
Metallurgy and manufacture of metal products	8,456	20	15,206	7
Machinery and equipment	7,507	22	14,929	11
Manufacture of machinery and equipment	7,917	35	14,973	14
Transport equipment	28,244	25	52,885	14
Food industry	−2,100,631	−25	−1,459,443	−15
Tobacco industry	−18,727	−15	−7,942	−5
Textiles, clothing and fur	8,958	7	24,616	8
Manufacture of leather and footwear	−19,507	−17	−1,757	−1
Paper industry	30,539	61	70,712	46
Wood, cork and wood furniture	22,139	98	49,976	256
Rubber, plastics and other manufactures	46,051	49	100,657	44
Construction and engineering	187,625	53	380,298	39
Recovery and Repair	0	–	0	–
Trade	−228,697	−18	−10,133	−1
Hotels and Restaurants	310,622	12	118,778	3
Transport and communications	30,965	32	47,313	7
Credit and insurance	4,049	17	7,793	11
Real estate	12,867	14	34,431	14
Private Education	−6,804	−11	−9,719	−8

(continued)

Table 3 (continued)

	Water withdrawals		Physical consumption (blue and green)	
	Difference in hm^3 (Agg-Disagg)	% wrt Disagg	Difference in hm^3 (Agg-Disagg)	% wrt Disagg
Private healthcare	−15,517	−16	−20,410	−10
Other services for sale	4,487	1	53,005	8
Domestic service	0	−	0	−
Public education	3,218	7	7,199	5
Public health	7,739	6	12,912	4
Public services	33,341	11	64,487	22
Total	0	0	0	0

Notes %wrt Disagg: % with respect to the disaggregated value
Source Own elaboration

In the case of hotels and restaurants final demand, 311 hm^3 (12%) more (hence opposite difference than for the case of the SAM for Huesca) is found with the aggregated table. We may see then something which we could not appreciate with the EXIOBASE database, which is how the detail in sectors strongly related with food, such as food trade and hotels and restaurants, which are also very close to distribution and final demand, matter enormously. Also, highly interestingly for us, although these activities have the largest absolute differences, also because as we had already explored, these activities of the agri-food system involve the largest embodied (virtual) water intensities, large differences, particularly percentage wise, are also found in other sectors. For example, the volumes in "Construction and engineering" final demand reveal noteworthy 187 hm^3 (53%) of water more is found with the aggregated table. Similarly, we may highlight the differences for "Wood, cork and wood furniture" (22 hm^3, 98% for water withdrawals, and 50 hm^3, representing an outstanding 256% of change with respect to the disaggregated figure), given that when forest sectors/products are considered together with other agricultural products (as in the aggregated version), the assumed water intensity (the average of the whole agricultural sector) becomes much larger than the real one. Other important cases are those of "Manufacture of leather and footwear" (for water withdrawals, −19 hm^3, −17%), "Paper industry" (31 hm^3, 61%), "Rubber, plastics and other manufactures" (46 hm^3, 49%), "Transport and communications" (31 hm^3, 32%), "Transport equipment" (28 hm^3, 25%). The main take home message from this finding then is that when we work with agri-food system aggregated sectors/products, clear biases/errors will be made for those sectors/products, but also for other related ones which sometimes only buy from them a few specific products.

5 Conclusions and Final Discussion

5.1 *Conclusions*

Different methods have been proposed and used for accounting virtual (or embodied) water and water footprints, and hence for water volumes in food products. This has to do with the question of truncation errors and the system boundaries selection, and to what extent the researcher prioritizes the specific analysis of the main supply chains (typically done with bottom-up methods) or to capturing larger supply chains, even when captured with much less precision. For water footprints the intuitive main "nodes" (sectors/products) and water flows can logically be thought to be the agri-food supply chain, and a few other related ones such as textiles, paper or bioenergy, which are the ones that the studies of the WFN have especially accounted for. The intuition or insight obviously is founded and grounded in data, notably of water uses and of intersectoral and international trade of products. For many and countries globally the main water user/consumer is agriculture (which in many IO tables is grouped all together in a single account, including also forestry), and in particular the crops production (about 60–70% of the water, depending on the metrics used). For this reason, it is probably fair to say that those type of studies are great in capturing the lion's share of water uses and flows in the world, furthermore with very detailed and relatively homogenous water intensities.

Our main inference is that the best approaches would be always taking as far as possible the best of "both worlds". In order to further approach the results for food products, and also possible biases, notably the sector aggregation ones, we performed two empirical applications, firstly from global databases, and secondly from an agri-food system disaggregated table for Spain. In the first type of studies, we highlighted that MRIO type of databases tend to have country detail, particularly for the richest countries. Among their multiple advantages, we may highlight the attempts of consistently reflecting trade, notably with the distinction of intermediate and final demand goods and services. Although we found quite distinct absolute figures of water uses among the most popular databases, given our purposes of exploring the specific agri-food supply chains and the relative sizes of aggregation we performed the estimates with EXIOBASE 2.2., revealing the ranking of activities embodying larger water footprints (hm^3) of consumption (WF C) for different water types (green, blue, blue "scarce", withdrawal, withdrawal "scarce"). We discovered that both by countries and by sectors, and both globally and at local levels, there is a high degree of heterogeneity, which reveals that any attempt at aggregation will frequently lead to serious errors, especially in those products-countries that move away from the aggregated intensity values used.

In the second type of tables and application, we also compared the results of the "disaggregated" (for Huesca/Spain) Social Accounting Matrices (SAMs) and the "aggregated" version packing it into 1 single agriculture category. In the case of the SAM for Spain, the aggregation also involves a category of the food sector, another of trade, and another of hotels and restaurants. We saw that in terms of absolute values, as one could expect, the largest differences were found for the aggregated sectors themselves, relatively logically because the agri-food system involves the largest embodied (virtual) water intensities.

In these final analyses we could see then something which we could not appreciate with the EXIOBASE database, which is how the detail in sectors strongly related with food, such as food trade and hotels and restaurants, which are also very close to distribution and final demand, matter enormously.

But also extremely interestingly for us, although these activities have the largest absolute differences, large differences, particularly percentage wise, are also found in other sectors. For example with the SAM for Spain, the volumes in "Construction and engineering" final demand reveal noteworthy 187 hm^3 (53%) of water more with the aggregated table, and "Wood, cork and wood furniture" (22 hm^3, 98% for water withdrawals, and 50 hm^3 of physical consumption, representing an outstanding 256% of change with respect to the disaggregated figure). Consequently, we may conclude that when we work with agri-food system aggregated sectors/products, clear biases/errors will be made for those sectors/products, but also for other related ones.

5.2 Further Discussion

The simple analyses performed could be framed as part of the studies on aggregation biases (see e.g. Ahmed and Wyckoff 2003; Peters et al. 2011; Weber 2008), but also on uncertainties of IO modelling (see among many Bullard and Sebald 1988; Hawkins et al. 2007; Hendrickson et al. 2006; Jackson and West 1989; Lenzen 2001; Morgan and Henrion 1990; Temurshoev 2015), and related research, so those insights should be also beard in mind.

The notable differences in water footprints which may be obtained under the different methods relates to the analogous discussion[4] raised by Kastner et al. (2014) and also analyzed e.g. in Bruckner et al. (2014, 2017) on the contradictory results from physical trade matrices (a kind of bottom-up type of study) and MRIO studies for the study of embodied (virtual) cropland. In that case, in our opinion part of the story could be explained by the fact that in the physical trade matrices a sector such as textiles and leather was not considered, involving large embodied land from the purchases of crops and livestock products (which required huge volumes of crops as feed as well). But certainly, the differences were so large that it seemed that some other factors need to play a role.

---

[4]Another similar discussion was that opened by Schoer et al. (2013) on the differences of life cycle inventory data versus MRIO for raw material equivalents embodied in EU27 imports.

Acknowledgements The authors acknowledge financial support from the projects ECO2016-74940-P and ECO2013-41353-P granted by the Spanish Ministry of Education and Science; as well as the financial support from Consolited Group S10 and Reference Group S40-17R granted by the Aragonese Government (DGA) and the European Regional Development Fund (ERDF).

References

Aguilar, S., Louw, K., & Neville, K. (2011). *IHA World Congress Bulletin.*

Ahmed, N., & Wyckoff, A. (2003). *Carbon dioxide emissions embodied in international trade of goods.*

Aldaya, M. M., Martínez-Santos, P., & Llamas, M. R. (2010). Incorporating the water footprint and virtual water into policy: Reflections from the Mancha Occidental Region, Spain. *Water Resources Management, 24,* 941–958. https://doi.org/10.1007/s11269-009-9480-8.

Allan, J. A. (1993). Fortunately there are substitutes for water otherwise our hydro-political futures would be impossible. *Priorities for Water Resources Allocation and Management,* 13–26.

Allan, J. A. (1994). Overall perspectives on countries and regions. *Water in the Arab World: Perspectives and Prognosis,* 65–100.

Allan, J. A. (1996). Policy responses to the closure of water resources: Regional and global issues. In P. Howsam & R. C. Carter (Eds.), *Water policy: Allocation and management in practice. London Proceedings of International Conference on Water Policy.* Cranfield University.

Allan, J. A. (1998). Virtual water: A strategic resource global solutions to regional deficits. *Ground Water, 36,* 545–546.

Bayart, J.-B., Bulle, C., Deschênes, L., Margni, M., Pfister, S., Vince, F., et al. (2010). A framework for assessing off-stream freshwater use in LCA. *The International Journal of Life Cycle Assessment, 15,* 439–453. https://doi.org/10.1007/s11367-010-0172-7.

Berger, M., & Finkbeiner, M. (2010). Water footprinting: How to address water use in life cycle assessment? *Sustainability, 2,* 919–944.

Berger, M., & Finkbeiner, M. (2012). Methodological challenges in volumetric and impact-oriented water footprints. *Journal of Industrial Ecology, 17,* 79–89. https://doi.org/10.1111/j.1530-9290.2012.00495.x.

Bogra, S., Bakshi, B. R., & Mathur, R. (2016). A water-withdrawal input-output model of the Indian Economy. *Environmental Science & Technology, 50,* 1313–1321. https://doi.org/10.1021/acs.est.5b03492.

Boulay, A.-M., Bare, J., Benini, L., Berger, M., Lathuillière, M.J., Manzardo, A., et al. (2018). The WULCA consensus characterization model for water scarcity footprints: Assessing impacts of water consumption based on available water remaining (AWARE). *The International Journal of Life Cycle Assessment, 23,* 368–378. https://doi.org/10.1007/s11367-017-1333-8.

Bouwmeester, M. C., & Oosterhaven, J. (2013). Specification and aggregation errors in environmentally extended input-output models. *Environmental and Resource Economics, 56,* 307–335. https://doi.org/10.1007/s10640-013-9649-8.

Bruckner, M., Giljum, S., Fischer, G., & Tramberend, S. (2014). *Review of land flow accounting methodologies and recommendations for further development* (pp. 1–58). https://doi.org/TEXTE77/2017.

Bruckner, M., Giljum, S., Fischer, G., & Tramberend, S. (2017). *Review of land flow accounting methods and recommendations for further development.* Welthandelsplatz 1, 1020, Vienna, Austria.

Bullard, C. W., & Sebald, A. V. (1988). Monte Carlo sensitivity analysis of input-output models. *The Review of Economics and Statistics, 70,* 708–712. https://doi.org/10.2307/1935838.

Cazcarro, I., & Arto, I. (2018). Water footprint and consumer products. In S. S. Muthu (Ed.), *Environmental footprints of water: Case studies from across industrial sectors*. Elsevier B.V.

Cazcarro, I., Pac, R. D., & Sánchez-Chóliz, J. (2010). Water consumption based on a disaggregated social accounting matrix of Huesca (Spain). *Journal of Industrial Ecology, 14.* https://doi.org/10.1111/j.1530-9290.2010.00230.x.

Cazcarro, I., Duarte, R., & Sánchez-Chóliz, J. (2012). Water flows in the Spanish economy: Agri-food sectors, trade and households diets in an input-output framework. *Environmental Science & Technology, 46,* 6530–6538. https://doi.org/10.1021/es203772v.

Cazcarro, I., Hoekstra, A. Y., & Sánchez Chóliz, J. (2014). The water footprint of tourism in Spain. *Tourism Management, 40.* https://doi.org/10.1016/j.tourman.2013.05.010.

Cazcarro, I., López-Morales, C. A. & Duchin, F. (2016). The global economic costs of the need to treat polluted water. *Economic Systems Research, 28*(3), 295–314.

Chapagain, A. K., & Hoekstra, A. Y. (2003). *Virtual water flows between nations in relation to trade in livestock and livestock products.* Netherlands: Delft.

Chapagain, A. K., & Hoekstra, A. Y. (2004). *Water footprints of nations, value of water.* Research Report Series.

Chapagain, A. K., Hoekstra, A. Y., Savenije, H. H. G., & Gautam, R. (2006). The water footprint of cotton consumption: An assessment of the impact of worldwide consumption of cotton products on the water resources in the cotton producing countries. *Ecological Economics, 60,* 186–203. https://doi.org/10.1016/j.ecolecon.2005.11.027.

Chenoweth, J., Hadjikakou, M., & Zoumides, C. (2014). Quantifying the human impact on water resources: A critical review of the water footprint concept. *Hydrology and Earth System Sciences, 18,* 2325–2342. https://doi.org/10.5194/hess-18-2325-2014.

Coca-Cola and Nature Conservancy. (2010). *Product water footprint assessments: Practical application in corporate water stewardship.*

Cooley, H., Fulton, J., & Gleick, P. (2011). *Water for energy: Future water needs for electricity in the Intermountain West.* USA: Oakland.

Cooper, T., Fallender, S., Pafumi, J., Dettling, J., Humbert, S., & Lessard, L. (2011). *A semiconductor company's examination of its water footprint approach.*

Crown, W. H. (1990). An interregional perspective on aggregation bias and information loss in input-output analysis. *Growth and Change, 21,* 11–29.

Daniels, P. L., Lenzen, M., & Kenway, S. J. (2011). The ins and outs of water use—A review of multi-region input–output analysis and water footprints for regional sustainability analysis and policy. *Economic Systems Research, 23,* 353–370. https://doi.org/10.1080/09535314.2011.633500.

de Koning, A., Bruckner, M., Lutter, S., Wood, R., Stadler, K., & Tukker, A. (2015). Effect of aggregation and disaggregation on embodied material use of products in input–output analysis. *Ecological Economics, 116,* 289–299. https://doi.org/10.1016/j.ecolecon.2015.05.008.

Dietzenbacher, E., Los, B., Stehrer, R., Timmer, M., & de Vries, G. (2013). The construction of world input–output tables in the WIOD project. *Economic Systems Research, 25,* 71–98. https://doi.org/10.1080/09535314.2012.761180.

Dominguez-Faus, R., Powers, S. E., Burken, J. G., & Alvarez, P. J. (2009). The water footprint of biofuels: A drink or drive issue? *Environmental Science & Technology, 43,* 3005–3010. https://doi.org/10.1021/es802162x.

Duarte, R., & Yang, H. (2011). Input–output and water: Introduction to the special issue. *Economic Systems Research, 23,* 341–351. https://doi.org/10.1080/09535314.2011.638277.

Ekvall, T., & Weidema, B. P. (2004). System boundaries and input data in consequential life cycle inventory analysis. *International Journal of Life Cycle Assessment, 9,* 161–171.

Eurostat. (2016). The Figaro project: The EU inter-country supply, use and input-output tables. In *Economic Commission for Europe. Conference of European Statisticians*. Group of Experts on National Accounts. Economic and Social Council, Geneva (pp. 1–169).

Feng, K., Chapagain, A., Suh, S., Pfister, S., & Hubacek, K. (2011). Comparison of bottom-up and top-down approaches to calculating the water footprints of nations. *Economic Systems Research, 23,* 371–385. https://doi.org/10.1080/09535314.2011.638276.

Ferrão, P., & Nhambiu, J. (2009). A comparison between conventional LCA and hybrid EIO-LCA: Analyzing crystal giftware contribution to global warming potential BT. In S. Suh (Ed.), *Handbook of input-output economics in industrial ecology* (pp. 219–230). Springer Netherlands, Dordrecht. https://doi.org/10.1007/978-1-4020-5737-3_11.

Flörke, M., Kynast, E., Bärlund I., Eisner, S., Wimmer, F., & Alcamo, J. (2013). Domestic and industrial water uses of the past 60 years as a mirror of socio-economic development: A global simulation study. *Global Environmental Change, 23*(1), 144–156. https://doi.org/10.1016/j.gloenvcha.2012.10.018.

Gawel, E., & Bernsen, K. (2011). Do we really need a water footprint? Global trade, water scarcity and the limited role of virtual water. *GAIA-Ecological Perspectives for Science and Society, 20,* 162–167.

Gawel, E., & Bernsen, K. (2013). What is wrong with virtual water trading? On the limitations of the virtual water concept. *Environment and Planning C: Government and Policy, 31,* 168–181.

Haqiqi, I., Taheripour, F., Liu, J., & van der Mensbrugghe, M. (2016). Introducing irrigation water into GTAP data base version 9. *Journal of Global Economic Analysis, 1*(2), 116–155. https://jgea.org/resources/jgea/ojs/index.php/jgea/article/view/35.

Hauke, W., Leonie, W., Steckel, J. C., & Minx, J. C. (2017). Truncation error estimates in process life cycle assessment using input-output analysis. *Journal of Industrial Ecology.* https://doi.org/10.1111/jiec.12655.

Hawkins, T., Hendrickson, C., Higgins, C., Matthews, H. S., & Suh, S. (2007). A mixed-unit input-output model for environmental life-cycle assessment and material flow analysis. *Environmental Science & Technology, 41,* 1024–1031. https://doi.org/10.1021/es060871u.

Hendrickson, C. T., Lave, L. B., Matthews, H. S., Horvath, A., Joshi, S., McMichael, F. C., et al. (2006). *Environmental life cycle assessment of goods and services. An input-output approach.* https://doi.org/10.4324/9781936331383.

Herath, I., Deurer, M., Horne, D., Singh, R., & Clothier, B. (2011). The water footprint of hydroelectricity: A methodological comparison from a case study in New Zealand. *Journal of Cleaner Production, 19,* 1582–1589. https://doi.org/10.1016/j.jclepro.2011.05.007.

Hoekstra, A. Y. (2013). Wise freshwater allocation: Water footprint caps by river basin, benchmarks by product and fair water footprint shares by community. *Value of Water Research Report BT - Wise freshwater allocation: Water footprint caps by river basin, benchmarks by product and fair water footprint shares by community.*

Hoekstra, A. Y. (2016). A critique on the water-scarcity weighted water footprint in LCA. *Ecological Indicators, 66,* 564–573. https://doi.org/10.1016/j.ecolind.2016.02.026.

Hoekstra, A. Y., & Chapagain, A. K. (2007). Water footprints of nations: Water use by people as a function of their consumption pattern. *Water Resources Management, 21,* 35–48. https://doi.org/10.1007/s11269-006-9039-x.

Hoekstra, A. Y., & Chapagain, A. K. (2008). *Globalization of water: Sharing the planet's freshwater resources.*

Hoekstra, A. Y., & Hung, P. Q. (2002). Virtual water trade: A quantification of virtual water flows between nations in relation to international crop trade. In *Virtual water trade a quantification virtual water flows between nations relation to International Crop Trade.*

Hoekstra, A. Y., Chapagain, A. K., Aldaya, M., & Mekonnen, M. M. (2009). *Water footprint manual: State of the art 2009, Water footprint network.* Enschede, The Netherlands.

Hoekstra, A. Y., Gerbens-Leenes, W., & Van Der Meer, T. H. (2009). Reply to Pfister and Hellweg: Water footprint accounting, impact assessment, and life-cycle assessment. *Proceedings of the National Academy of Sciences of the United States of America, 106.* https://doi.org/10.1073/pnas.0909948106.

Hoekstra, A. Y., Chapagain, A. K., Aldaya, M., Mekonnen, M. M., & Chapagain, A. K. (2011). *The water footprint assessment manual: Setting the global standard.* London, UK.

Humbert, S. (2009). *ISO standard on water footprint: Principles, requirements and guidance.*

ISO. (2014). *ISO briefing note: Measuring the impact of water use and promoting efficiency in water management.*

ISO. (2017). *Environmental management—Water footprint—Illustrative examples on how to apply ISO 14046*.

Jackson, R. W., & West, G. R. (1989). Perspectives on probabilistic input-output analysis. In R. Miller, K. Polenske, & A. Rose (Eds.), *Frontiers of input-output analysis*. London: Oxford University Press.

Jalili, A. R. (2005). Impacts of aggregation on relative performances of nonsurvey updating techniques and intertemporal stability of input-output coefficients. *Economic Change and Restructuring, 38,* 147–165.

Jesper, M., Mette, W., Manfred, L., & Christopher, D. (2008). Using input-output analysis to measure the environmental pressure of consumption at different spatial levels. *Journal of Industrial Ecology, 9,* 169–185. https://doi.org/10.1162/1088198054084699.

Kastner, T., Schaffartzik, A., Eisenmenger, N., Erb, K.-H., Haberl, H., & Krausmann, F. (2014). Cropland area embodied in international trade: Contradictory results from different approaches. *Ecological Economics, 104,* 140–144. https://doi.org/10.1016/j.ecolecon.2013.12.003.

Kounina, A., Margni, M., Bayart, J.-B., Boulay, A.-M., Berger, M., Bulle, C., et al. (2013). Review of methods addressing freshwater use in life cycle inventory and impact assessment. *The International Journal of Life Cycle Assessment, 18,* 707–721. https://doi.org/10.1007/s11367-012-0519-3.

Kymn, K. O. (1990). Aggregation in input-output models: a comprehensive review, 1946–71. *Economic Systems Research, 2,* 65–93.

Lahr, M. L., & Stevens, B. H. (2002). A study of the role of regionalization in the generation of aggregation error in regional input-output models. *Journal of Regulatory Science, 42,* 477–507. https://doi.org/10.1111/1467-9787.00268.

Lenzen, M. (2001). Errors in conventional and input-output-based life-cycle inventories. *Journal of Industrial Ecology, 4,* 127–148. https://doi.org/10.1162/10881980052541981.

Lenzen, M. (2009). Understanding virtual water flows: A multiregion input-output case study of Victoria. *Water Resources Research, 45,* 1–11. https://doi.org/10.1029/2008WR007649.

Lenzen, M. (2011). Aggregation versus disaggregation in input–output analysis of the environment. *Economic Systems Research, 23,* 73–89. https://doi.org/10.1080/09535314.2010.548793.

Lenzen, M., & Foran, B. (2001). An input-output analysis of Australian water usage. *Water Policy, 3,* 321–340. https://doi.org/10.1016/s1366-7017(01)00072-1.

Lenzen, M., & Treloar, G. (2002). Embodied energy in buildings: Wood versus concrete—Reply to Börjesson and Gustavsson. *Energy Policy, 30,* 249–255. https://doi.org/10.1016/s0301-4215(01)00142-2.

Lenzen, M., & Treloar, G. (2003). Differential convergence of life-cycle inventories toward upstream production layers: Implications for life-cycle assessment. *Journal of Industrial Ecology, 6,* 137–160.

Lenzen, M., Pade, L. L., & Munksgaard, J. (2004). CO_2 multipliers in multi-region input-output models. *Economic Systems Research, 16,* 389–412. https://doi.org/10.1080/0953531042000304272.

Lenzen, M., Bhaduri, A., Moran, D., Kanemoto, K., Bekchanov, G. A., Foran, B., et al. (2012a). *The role of scarcity in global virtual water flows*. ZEF-Discussion Papers on Development Policy 24.

Lenzen, M., Bhaduri, A., Moran, D., & Kanemoto, K. (2012b). *The role of scarcity in global virtual water flows*. ZEF-Discussion Papers on Development Policy No. 169

Lenzen, M., Kanemoto, K., Moran, D., & Geschke, A. (2012c). Mapping the structure of the world economy. *Environmental Science & Technology, 46,* 8374–8381. https://doi.org/10.1021/es300171x.

Lenzen, M., Moran, D., Bhaduri, A., Kanemoto, K., Bekchanov, M., Geschke, A., et al. (2013a). International trade of scarce water. *Ecological Economics, 94,* 78–85. https://doi.org/10.1016/j.ecolecon.2013.06.018.

Lenzen, M., Moran, D., & Kanemoto, K. (2013b). Building EORA: A global multi-region input–output database at high country and sector resolution. *Economic Systems Research, 25,* 37–41.

Lifset, R. (2009). Industrial ecology in the age of input-output analysis BT. In S. Suh (Ed.), *Handbook of input-output economics in industrial ecology* (pp. 3–21). Springer Netherlands, Dordrecht. https://doi.org/10.1007/978-1-4020-5737-3_1.

Lutter, S., & Giljum, S. (2015). *Developing targets for global resource use targets for freshwater use*.

Lutter, S., Mekonnen, M. M., & Raptis, C. (2013). Updated and improved data on water consumption/use imported into the EXIOBASE in the required sectoral (dis)aggregation. *CREEA Deliv., 3, 4.*

Lutter, S., Giljum, S., Pfister, S., Raptis, C., Mutel, C., & Mekonnen, M. M. (2014). D8.1 - CREEA Water Case Study Report. Vienna/Zurich/Twente.

Mekonnen, M. M., & Hoekstra, A. Y. (2010). *The green, blue and grey water footprint of farm animals and animal products* (Vol. 1). https://doi.org/10.5194/hess-15-1577-2011.

Mekonnen, M. M., & Hoekstra, A. Y. (2011a). The green, blue and grey water footprint of crops and derived crop products. *Hydrology and Earth System Sciences, 15*, 1577–1600. https://doi.org/10.5194/hess-15-1577-2011.

Mekonnen, M. M., & Hoekstra, A. Y. (2011b). *The blue water footprint of electricity from hydropower.*

Mekonnen, M. M., & Hoekstra, A. Y. (2014). Water footprint benchmarks for crop production: A first global assessment. *Ecological Indicators, 46*, 214–223. https://doi.org/10.1016/j.ecolind.2014.06.013.

Milà i Canals, L., Chenoweth, J., Chapagain, A., Orr, S., Antón, A., Clift, R., et al. (2009). Assessing freshwater use impacts in LCA: Part I—Inventory modelling and characterisation factors for the main impact pathways. *The International Journal of Life Cycle Assessment, 14*, 28–42. https://doi.org/10.1007/s11367-008-0030-z.

Miller, R. E., & Blair, P. D. (2009). *Input-output analysis.*

Morgan, G., & Henrion, M. (1990). *Uncertainty.* Cambridge: Cambridge University Press.

Morimoto, Y. (1970). On aggregation problems in input-output analysis. *The Review of Economic Studies, 37*, 119–126. https://doi.org/10.2307/2296502.

Morimoto, Y. (1971). A note on weighted aggregation in input-output analysis. *International Economic Review (Philadelphia), 12*, 138–143. https://doi.org/10.2307/2525502.

Narayanan, G., Aguiar, A., & McDougall, R. (2012). *Global trade, assistance, and production: The GTAP 8 Data Base, Center for Global Trade Analysis.*

Narayanan, G., Badri, A. A., & McDougall, R. (2015). *Global trade, assistance, and production: The GTAP 9 Data Base, Center for Global Trade Analysis.* Purdue University.

OECD. (2016). *OECD Inter-Country Input-Output (ICIO).*

Olsen, J. A. (2000). Aggregation in macroeconomic models: An empirical input-output approach. *Economic Modelling, 17*, 545–558.

Pepsi-Co. (2011). *Environmental Sustainability Report 2009/10: Path to Zero.*

Perry, C. (2014). Water footprints: Path to enlightenment, or false trail? *Agricultural Water Management, 134*, 119–125.

Peters, G. P., Minx, J. C., Weber, C. L., & Edenhofer, O. (2011). Growth in emission transfers via international trade from 1990 to 2008. *Proceedings of the National Academy of Sciences, 108*, 8903–8908. https://doi.org/10.1073/pnas.1006388108.

Pfister, S., Bayer, P., Koehler, A., & Hellweg, S. (2011). Environmental impacts of water use in global crop production: Hotspots and trade-offs with land use. *Environmental Science & Technology, 45*(13), 5761–5768. https://doi.org/10.1021/es1041755.

Pfister, S., & Bayer, P. (2014). Monthly water stress: Spatially and temporally explicit consumptive water footprint of global crop production. *Journal of Cleaner Production, 73*, 52–62. https://doi.org/10.1016/j.jclepro.2013.11.031.

Pfister, S., Boulay, A.-M. M., Berger, M., Hadjikakou, M., Motoshita, M., Hess, T., et al. (2017). Understanding the LCA and ISO water footprint: A response to Hoekstra (2016) "A critique on the water-scarcity weighted water footprint in LCA." *Ecological Indicators, 72*, 352–359. https://doi.org/10.1016/j.ecolind.2016.07.051.

Pittel, K., Rübbelke, D. D., & Altemeyer-Bartscher, M. (2012). International efforts to combat global warming. In W.-Y. Chen, J. Seiner, T. Suzuki, & M. Lackner (Eds.), *Handbook of climate change mitigation* (pp. 89–120). US: Springer.

Postel, S. L. (2000). Entering an era of water scarcity: The challenges ahead. *Ecological Applications, 10*, 941–948.

Ridoutt, B. G., & Pfister, S. (2010). A revised approach to water footprinting to make transparent the impacts of consumption and production on global freshwater scarcity. *Global Environmental Change, 20*, 113–120. https://doi.org/10.1016/j.gloenvcha.2009.08.003.

SABMiller_WWF-UK. (2009). *Water footprinting: Identifying and addressing water risks in the value chain*. Woking.

Schoer, K., Wood, R., Arto, I., & Weinzettel, J. (2013). Estimating raw material equivalents on a macro-level: Comparison of multi-regional input–output analysis and hybrid LCI-IO. *Environmental Science & Technology, 47*, 14282–14289. https://doi.org/10.1021/es404166f.

Stadler, K., Wood, R., & Bulavskaya, T. (2016). *Development of EXIOBASE 3 EXIOBASE database framework*.

Stadler, et al. (2018). EXIOBASE 3: Developing a time series of detailed environmentally extended multi-regional input-output tables: EXIOBASE 3. *Journal of Industrial Ecology, 22* (3), 502–515. https://onlinelibrary.wiley.com/doi/pdf/10.1111/jiec.12715.

Stehrer, R., & Rueda-Cantuche, J. M. (2016). *Review of the methods for the estimation of global multi-country supply, use and input-output tables*.

Su, B., & Ang, B. W. (2010). Input–output analysis of CO_2 emissions embodied in trade: The effects of spatial aggregation. *Ecological Economics, 70*, 10–18. https://doi.org/10.1016/j. ecolecon.2010.08.016.

Suh, S. (2004). Functions, commodities and environmental impacts in an ecological-economic model. *Ecological Economics, 48*, 451–467. https://doi.org/10.1016/j.ecolecon.2003.10.013.

Suh, S., Lenzen, M., Treloar, G. J., Hondo, H., Horvath, A., Huppes, G., et al. (2004). System boundary selection in life-cycle inventories using hybrid approaches. *Environmental Science & Technology, 38*, 657–664. https://doi.org/10.1021/es0263745.

Suh, S., Wirtschaftsuniversit, S. G., & Barbara, S. (2009). *Handbook of input-output economics in industrial ecology, eco-efficiency in industry and science*. Springer. https://doi.org/10.1007/ 978-1-4020-5737-3

Taheripour, F., Hertel, T., & Liu, J. (2013). Introducing water by river basin into the GTAP-BIO model: GTAP-BIO-W. Department of Agricultural Economics, Purdue University, West Lafayette, IN: Global Trade Analysis Project (GTAP). https://www.gtap.agecon.purdue.edu/ resources/res_display.asp?RecordID=4304.

Temurshoev, U. (2015). Uncertainty treatment in input-output analysis. In *Handbook of input-output analysis* (pp. 407–463). https://doi.org/10.4337/9781783476329.00018.

Tol, R. S. J. (2011). *The GTAP-W model: Accounting for Water Use in Agriculture by Alvaro Calzadilla*, Katrin Rehdanz, no. 1745.

Tol, R. S. J., Alvaro, C., Katrin, R., & Tol Richard, S. J. (2011). *The GTAP-W model: Accounting for water use in agriculture*, no. 1745. http://ideas.repec.org/p/kie/kieliw/1745.html.

Treloar, G. J. (1997). Extracting embodied energy paths from input-output tables: Towards an input-output-based hybrid energy analysis method. *Economic Systems Research, 9*, 375–391.

Tukker, A., Poliakov, E., Heijungs, R., Hawkins, T., Neuwahl, F., Rueda-Cantouche, J., et al. (2009). Towards a global multiregional environmentally-extended input-output database. *Ecological Economics, 68*, 1928–1937.

Tukker, A., de Koning, A., Wood, R., Hawkins, T., Lutter, S., Acosta, J., et al. (2013). EXIOPOL —Development and illustrative analyses of a detailed global MR EE SUT/IOT. *Economic Systems Research, 25*, 50–70. https://doi.org/10.1080/09535314.2012.761952.

UN_Comtrade. (2018). *United Nations commodity trade statistics database*.

Vanham, D., & Bidoglio, G. (2013). A review on the indicator water footprint for the EU28. *Ecological Indicators, 26*, 61–75. https://doi.org/10.1016/j.ecolind.2012.10.021.

Verma, S., Kampman, D.A., Van der Zaag, P., & Hoekstra, A. Y. (2009). Going against the flow: A critical analysis of inter-state virtual water trade in the context of India's National River Linking Program. *Value* of *Water Research Report Series, 34*, 261–269. https://doi.org/10. 1016/j.pce.2008.05.002.

Vörösmarty, C. J., Hoekstra, A. Y., Bunn, S. E., Conway, D., & Gupta, J. (2015). Fresh water goes global. *Science, 349* (80–.), 478–479.

Weber, C. L. (2008). Uncertainties in constructing environmental multiregional input-output models. In *International input–output meeting on managing the environment*

WFN. (2018). *Water Footprint Network [WWW Document].* http://www.waterfootprint.org.

Wichelns, D. (2010). Virtual water: A helpful perspective, but not a sufficient policy criterion. *Water Resources Management, 24*, 2203–2219. https://doi.org/10.1007/s11269-009-9547-6.

Wichelns, D. (2011a). Footprint perspectives enhance policy discussions? Do the virtual water and water footprint perspectives enhance policy discussions? (pp. 37–41).

Wichelns, D. (2011b). Virtual water and water footprints: compelling notions, but notably flawed: Reaction to two articles regard the virtual water concept. *GAIA, 20*, 171–175.

Wichelns, D. (2015). Virtual water and water footprints: Overreaching into the discourse on sustainability, efficiency, and equity. *Water Alternatives, 8*, 396–414.

Witmer, M., & Cleij, P. (2012). *Water Footprint: Useful for sustainability policies?* The Hague: PBL Netherlands Environmental Agency. https://doi.org/PBL. Publication number: 500007001.

Wood, R., Stadler, K., Bulavskaya, T., Lutter, S., Giljum, S., de Koning, A., et al. (2015). Global sustainability accounting—Developing EXIOBASE for multi-regional footprint analysis. *Sustainability.* https://doi.org/10.3390/su7010138.

WWF. (2012). *Living planet report 2012.* Switzerland: Gland.

Yang, H., Pfister, S., & Bhaduri, A. (2013). Accounting for a scarce resource: Virtual water and water footprint in the global water system. *Current Opinion in Environmental Sustainability, 5*, 599–606. https://doi.org/10.1016/j.cosust.2013.10.003.

Zhang, Y., Huang, K., Yu, Y., & Yang, B. (2017). Mapping of water footprint research: A bibliometric analysis during 2006–2015. *Journal of Cleaner Production, 149*, 70–79. https://doi.org/10.1016/j.jclepro.2017.02.067.

Printed in the United States
By Bookmasters